U0364179

酒店+ Hotel

Luxury Style 奢华风

2

佳图文化　主编

中国林业出版社

Contents

目录

Resort Hotel 度假酒店

Business Hotel 商务酒店

百乐酒店加大扩张步伐　今年计划推出新品牌

源于新加坡的百乐酒店集团过去一直以自持酒店项目为主要发展模式，今年起集团为扩大在亚太地区的知名度与版图，将重心朝向“酒店管理”项目发展，并计划最快在今年内推出继百乐及君乐品牌之外的第三个全新品牌。百乐集团日前刚成功签下两家位于新加坡的酒店管理项目，一家是位于花拉公园地铁站上方的新加坡百乐景园酒店、另一家坐落于新加坡亚历山大路以及惹兰红山交界处的新加坡百乐亚历山大酒店，两者最快都将于 2015 年开业。

Park Hotel Group Looks for Great Expansion: Launch New Brand in 2013

With its first hotel opened in Singapore, Park Hotel Group is insisting on developing self-owned hotels in these years. And from this year, it will expand its brand and hotel assets in the Asian-Pacific region with “hotel management” as its new focus. It plans to present a new brand in addition to Grand Park and Park Hotel. The group has just signed two hotel management projects in Singapore. One is the Park Hotel Farrer Park on top of the Farrer Park MRT station. The other is the Park Hotel Alexander located on Alexander Road.Two of them are expected to be completed and opened in 2015.

丽思卡尔顿酒店集团新酒店落户天津历史核心区

天津丽思卡尔顿酒店位于前英租界风貌保护区——泰安道英式风情区，是城市中央商务区的核心地带。酒店作为一个新的地标性建筑，被视为对这座城市的历史遗产和现代发展的敬意，将于 2013 年中期开业。酒店欧洲古堡式的建筑由天津市城市规划设计研究院精心打造。内部装饰由知名的法国奢华酒店设计师 Pierre-Yves Rochon 主创设计，新古典主义风格体现在其中。酒店 277 间客房舒适豪华，其中包括 53 间套房，都彰显着该区域欧洲传统文化的高贵雅致。

The Ritz-Carlton Hotel Company L.L.C. Anchors Northeast China with 2013 Opening in Revitalized Historical Heart of Tianjin

Located in the heart of the revitalized former British concession and central business district, The Ritz-Carlton, Tianjin will form an iconic landmark when it opens mid-2013 and pays homage to the city's historical heritage and vibrant modern development. The soon-to-open Ritz-Carlton hotel with a grand neo-classical style façade was designed by the Tianjin Academy of Urban Planning and Design, with interiors created by luxury hospitality designer, Pierre-Yves Rochon. It houses 277 rooms, including 53 suites and reflects the cultural European heritage in the district.

国际超五星级万达文华酒店将于 8 月开业

天津万达文华酒店将于 8 月 30 日开业，开业后的万达文华酒店将为国际高端客人提供品牌服务。据悉，万达文华酒店是国际超五星级酒店，也是万达酒店及度假村管理有限公司在华北地区投资并管理的首家万达文华品牌酒店，开业后，必将为天津的旅游饭店业注入“个性、精致、愉悦”全新奢华的活力。

Wanda Vista Tianjin to Open in August

The super five-star Wanda Vista Tianjin is going to open on August 30, 2013. It will provide international customers with high-end services. It is the first Wanda Vista Hotel in Northern China that invested and managed by Wanda Hotels & Resorts Co., Ltd. Its opening will bring the hotel industry of Tianjin with “personalized, boutique and pleasant” luxury experience.

2013 酒店营销高峰论坛将于 5 月 15 日举行

5 月 15 日 ~16 日，环球旅讯主办的“中国酒店营销高峰论坛”将在上海东锦江希尔顿逸林酒店举行。论坛为期两天，通过主题演讲、嘉宾座谈、CEO 对话等形式，分析和探讨全球酒店的在线营销新趋势、酒店在线营销策略、新兴模式对酒店收益及渠道管理带来的机会和挑战，以及如何利用社会媒体和用户点评来提升酒店销售、分销和收益管理的绩效等精彩议题。届时将有超过 500 位来自全球酒店业的高层出席此次盛会，锦江酒店集团 CEOBernoldO.Schroeder、开元酒店集团总经理陈妙强、香格里拉酒店集团市场营销副总裁陈延菁、7 天连锁酒店副总裁李婉丽、速 8 酒店集团营销高级副总裁徐钊、携程旅行网高级副总裁孙茂华、艺龙旅行网 COO 谢震、Booking.com 缤客大中国区总经理赖军等 40 余位酒店营销业高层受邀将在此次“中国酒店营销高峰论坛”演讲或座谈，与业界精英分享其对酒店营销的独到观点和经验。

2013 China Hotel Marketing Summit to Be Held on May 15

China Hotel Marketing Summit is organized by TravelDaily. “Unbound world” is the main theme of China Hotel Marketing Summit 2013. Topics that will be covered include: Chinese Hotel Market Overview; How Hotels Leverage Technology and Service Innovations to Provide Better Booking and Stay Experience to Business Travelers and Improve Customer Loyalty; How the Hotels Leverage Innovative Digital Tools to Enhance Customer Loyalty and Improve Online Conversion; Hotel SOLOMO Strategy; etc. Over 500 C-level executives and senior management leaders from hotel industry will take center stage at the China Hotel Marketing Summit from May 15 to 16, 2013 in Double Tree by Hilton, Shanghai-Pudong to address the core issues affecting the heart of hotel industry.

三亚启动 2013 年度高端酒店评选活动

4 月 26 日，《新旅行》2013 年度高端酒店评选活动在三亚启动，来自国内外五星级高端酒店代表应邀出席了当天的新闻发布会。据悉，本年度高端酒店评选范围包括全国及亚太地区五星级及以上的奢华酒店，活动启动后，将经过投票和专家评审等多个环节，年底公布结果并举办颁奖典礼。

Best Hotel & Resort Value Award 2013 Started in Sanya

On April 26, Voyage magazine started the "Best Hotel & Resort Value Award 2013" and representatives of five-star hotels from all over the world were invited to the press conference. It is said that the entry hotels including the luxurious five-star hotels from China and Asian-Pacific region. The result will be launched at the end of this year after vote and expert review.

成都仁恒辉盛阁国际公寓获评“中国最佳国际酒店式服务公寓”

在第十三届中国饭店金马奖颁奖典礼上，仁恒辉盛阁国际公寓从中国酒店百强中脱颖而出，荣膺“第十三届中国饭店金马奖”，并被世界旅游组织亚太酒店协会授予“中国最佳国际酒店式服务公寓”称号。中国饭店金马奖由《中国饭店》杂志社联合世界旅游组织亚太酒店协会、中国社科院旅游研究中心等单位主办，是中国饭店业最高荣誉之一。中国饭店金马奖被视为行业发展的风向标，获奖企业和个人被标榜为行业典范。

Fraser Suites Chengdu Awarded "Best International Service Apartment in China"

Fraser Suites Chengdu won the 13th China Hotel Golden Horse Award and was titled as the "Best International Serviced Apartment in China" by Asia Pacific Hotel Association of World Tourism Organization (WTO). "China Hotel Golden Horse Award" is jointly organized by China Hotel magazine, Asia Pacific Hotel Association of WTO, Tourism Research Center of Chinese Academy of Social Science, etc. This award is the industry recognized "Oscar" of China hotel industry. It is not only a ceremony granting awards symbolizing industry highest honor, but also a whole sorting and checking of the development trend of national hotel industry in the year.

首家巧克力主题酒店：房内有巧克力喷泉

据外媒报道，全球第一间以巧克力为主题的酒店在英国伯恩茅斯开幕。这间酒店拥有 13 间以巧克力颜色及图像打造的主题房间，并且每间都配有巧克力喷泉，房客可以尽情享用。酒店酒吧里还提供巧克力鸡尾酒。在这里入住的每位房客，进入客房前都会先被引导至厨房，动手 DIY 制作巧克力，不管是制作黑巧克力、白巧克力或牛奶，旅客都可以随意发挥创意。

First Chocolate Boutique Hotel: Chocolate Fountain in Room

According to the foreign medias, the first chocolate themed hotel – The Chocolate Boutique Hotel was opened in Bournemouth of UK. As the only chocolate themed hotel in the world, it has 13 unique chocolaty bedrooms which provide chocolate fountains. The Chocolate Bar provides 'tantalizing' chocolate cocktail. Every guest will be introduced to the kitchen and make one's own chocolate innovatively.

深圳大梅沙京基喜来登酒店任命荣柯林为总经理

日前，大梅沙京基喜来登度假酒店任命荣柯林(Colin Vickers) 为酒店总经理。在此之前，荣柯林担任杭州西溪喜来登度假酒店的总经理职务。荣柯林来自英国，拥有超过 20 年酒店从业经验，他于 1996 年加入喜达屋酒店及度假村国际集团，先后于欧洲及亚洲多个国家和地区的著名国际品牌酒店担任高级管理职位，积累了丰富的酒店管理经验。

New GM Appointed at SHERATON Dameisha

SHERATON Dameisha Resort welcomed its new general manager, Colin Vickers, in March. Vickers has more than 20 years of experience in the hospitality industry and began his career with Starwood in 1996 as an assistant manager at Sheraton Park Tower Hotel, London. He then held different posts in several countries in Europe and Asia, and most recently, he was the general manager of Sheraton Hangzhou Wetland Park Resort, where he was in charge during the opening of the hotel.

凯悦酒店第一季度净利润同比下滑 20%

凯悦酒店 (H) 近日发布了 2013 财年第一季度财报。报告显示，凯悦酒店第一季度净利润为 800 万美元，比去年同期的 1 000 万美元下滑 20%。不计入某些一次性项目，凯悦酒店第一季度调整后净利润为 1 400 万美元，高于去年同期的 500 万美元；调整后每股收益为 9 美分，高于去年同期的 3 美分。凯悦酒店第一季度营收为 9.75 亿美元，比去年同期增长 1.8%，凯悦酒店第一季度营收比上一季度的 10 亿美元下滑 2.5%，每股收益比上一季度的 20 美分下滑 55%。

Hyatt Reports First Quarter 2013 Results: Net Gains Decreased by 20%

On May. 1, 2013, Hyatt Hotels Corporation ("Hyatt") (NYSE: H) reported its first quarter 2013 financial results. The report shows that the group revenue was $8 million in the first quarter of 2013 compared to $10 million in the first quarter of 2012, a decrease of 20%. Adjusted for special items, net income attributable to Hyatt was $14 million, or $0.09 per share, during the first quarter of 2013 compared to net income attributable to Hyatt of $5 million, or 0.03 per share, during the first quarter of 2013. The total revenues of Hyatt is $0.975 billion in the first quarter of 2013, an increase of 1.8% compared to the first quarter of 2012; a decrease of 2.5% compared to $1 billion in the fourth quarter of 2012, or decreased by55% compared to $0.2 per share in that quarter.

"绿色"二字倡导的是一种新风尚，一种时代趋向的新生活方式。酒店业界的绿色之举比比皆是，然而共性的一点就是：以绿色冠名，表现形式却各有千秋。倘若与酒店品牌的其它元素相较，酒店的绿色生态设计则有着独特的亮点，也是打造酒店品牌的重要途径。

为酒店品牌裁剪绿色新装

而今的酒店，早已不是以往住行旅馆所代替步途中转的客栈、驿站，随着社会压力和生活压力的增大，人们对自然舒适环境的渴求度越来越强，长期处于压抑的水泥建筑群里的人们也常常会向往自由的田园山林美景，因此绿色生态酒店很快成为了国际酒店业的潮流，随着社会的发展具有丰富的内容和多种类别。今天不论是在美洲、亚洲还是欧洲，"绿色生态酒店"都已经成为度假生活方式的最时髦词汇。

所谓绿色生态酒店就是以生态学原理为指导，用空间轮廓、群体组合、造型与色彩等各种景观设计手法并综合周围环境，为客人设计出的以绿色、生态、舒适为主题，集休闲娱乐、商务宴请、旅游度假于一体的休闲场所。生态绿色酒店因其优美舒适的景色风光、对生态环境的保护理念、有益于顾客身心健康的好处，很好地满足了人们在日常中亲近自然的愿望。

地域性：实现绿色生态不可或缺的手段

绿色生态酒店理念强调酒店从选址开始就把人工对周围生态系统的破坏减少到最小，还把降低能源消耗、支持当地的环境保护和社区发展作为整个运营的一个部分。在建筑生态性的要求下，地域性成为实现绿色生态酒店不可或缺的手段。地域性手法注重生态发展，选址不是改变基地环境，而是顺应地形地貌，充分利用自然环境因素；建筑技术经济条件扎根于地区的现实生活，常就地取材，方便使用过后还原于生态环境，充分考虑不同地域的文化特征与风土人情。因此地域性本身就体现了一种朴素的生态观念。比如位于新西兰东部著名环境保护地凯库拉 (Kaikoura) 的哈普库树顶屋酒店 (Hapuku Lodge) 就是这样。该酒店全部利用当地茂密的丛林，屋顶全为树木，掩映在森林中，与大片森林和谐地融为一体，从而体现出酒店与森林皆为一个整体的理念。斐济的库塞斯特度假村 (Jean-Michel Cousteau's Resort) 除了充分地利用热带阳光，建造成草屋酒店外，更加充分地利用了酒店所在区域"堪称世界上最好的海洋生物观赏地"的特殊属性，开发潜水观赏海洋生物项目，达到了人与自然和谐相处，促进相互认识的美好愿景。同时满足了人类天人合一和知行合一的强烈愿望。被誉为"全世界第一家被承认的黄金级环保建筑，并成为世界性环保酒店的典范之一"。

串连建筑与地方的文化生态关系

广义的生态概念不仅包括自然环境，也包括人工环境和人类的历史文化环境。建筑与地方历史文化的和谐一致，即文化生态，也是绿色生态酒店设计的内容之一。文化生态涉及的是建筑与地方文化的关系，即对地方历史文化的尊重。一方面，绿色生态酒店不能仅仅体现本地文化，同时应该与其历史环境、文化背景等保持整体的联系，这样才能使历史环境及地区的文脉得到进一步的加强和保护而不是被破坏和削弱；另一方面，绿色生态酒店所处的地域文化环境也为酒店增加了当地传统文化特色，使酒店成为与地域文化环境相结合的独一无二的建筑，而这种独特性正是现代度假酒店的一种重要的发展方向。在建筑设计中，主要通过以下手法实现：一是采用传统建筑形式，这里所指采用传统建筑形式并不是让新建的酒店建筑模仿其周围的历史建筑，而是在利用传统建筑形式的基础上，有一定的创新以适应现代功能和结构；二是将历史符号抽象简化并加以提炼，汲取旧建筑形式的艺术特征和精神实质，融入新形式之中并加以表现，形成经过提炼、抽象和升华了的新语言并最终创造出与老形式具有同样视觉效应的新形式，以此体现文化的延续。

成功的例子很多，如宁波五龙潭山野间的山居酒店，借鉴中国传统的山水筑园意趣，使建筑以低调的方式参与到自然环境之中。恰好反映出设计者为创作出满意的场所，从空间层次安排、空间序列安排、空间变化安排。空间过渡安排人手，在尊重原有场地和地形的原则下，不断通过空间整合手段，通过现代技术与传统文化的融合，达到丰富的地域表现的技能。

又如位于奥地利东南的布鲁毛罗格纳温泉酒店，它的座右铭是“微妙的不同”，它带给房客的视觉震撼力却实在是“大不同”。酒店有如一个小城镇，分成不同的公寓区，蜿蜒婉转的路，将你引向各种意想不到的景观。奥地利建筑艺术家百水先生（Hundert wasser）的风格在这里发挥得淋漓尽致：一组公寓群，竟将房舍嵌进了地下，如同一座座土拨鼠的地穴；别墅群的顶端种植着绿油油的草坪和灌木，甚至连窗户里都会探出一棵棵的树来，令你对建筑的认识被彻底颠覆。从温泉酒店的游泳池一路游到温泉池，不规则的水路忽左忽右、忽上忽下，眼前的景致一会儿是波斯宫殿，一会儿又变成了斑马小屋，令你忘记人世间所有的游戏规则。

地域性体现了一种朴素的生态观念。地域性手法注重生态发展，选址不是改变基地环境，而是顺应地形地貌；建筑技术经济条件扎根于地区的现实生活，充分考虑不同地域的文化特征与风土人情。

因地制宜的生态技术

绿色生态酒店的建筑设计，应始终突出休闲性特征，以实用为原则，避免大而全。常用的生态技术包括自然采光通风、遮阳技术，生态节水、节能减排技术等。

适宜技术也称中间技术或中等技术，它提倡的是技术的因地制宜性。这种中间技术与土技术（处于衰退的状况）相比，生产效率高得多，与现代工业的资本高度密集的高级技术相比又便宜得多。从绿色生态酒店设计地域性的角度来看，适宜建筑技术是针对地理、气候、地域文化的技术手段。它继承并发展传统，批判地吸收外来先进的实用技术。建筑师可以利用当地随地可取的材料，应用当地传统的手艺来施工。从设计层面来说，建筑师通过对现代材料与当地传统材料的对比与统一关系的把握，以现代的手法将两者有机巧妙地结合，促成建筑物与周围环境统一，同时也符合了成本上的要求。印尼巴厘岛的 Alfla Viuas Uluwatu 酒店设计方案是尊重保护环境的典型例子，其总体的规划尊重既有地形，避免挖土和填土。大树要么被保留要么移走。基地上的花圃培育本地的植物用于景观，而不是使用外来的植物。促进本地的动物和鸟类留在该地区，维持生态圈的平衡。该酒店从设计初始到最终提供产品所涉及的环境行为贯彻环保要求，建筑材料均采用了绿色的无污染材料。

创造与自然环境沟通的条件

绿色生态酒店作为联系使用者与自然环境的桥梁，应尽可能多地将自然元素引入到使用者身边，这也是生态原则的一个重要体现。在这里，建筑不再是隔绝人类与自然环境的厚重屏障，它将给人们提供一种崭新的生活。建筑中充满空气气息，充盈着柔和的日光而非各种荧光灯管。引入自然元素，为使用者与自然环境的沟通创造条件是生态度假酒店追求的另一个主要目标。引入自然元素的生态环境设计离不开三要素：植物、水景、山石。

在绿色生态环境设计中植物是必不可少的。绿色生态酒店设计中常通过各类植物的合理搭配，创造出景致各异的景观，愉悦人们的身心。其不但具有美化环境、陶冶情操的功能，还具有改善环境、净化空气的作用。植物再搭配山水则成生态环境。生态酒店中植物的设计原则：考虑到生态酒店内部植物生长采光需要，在邻近生态酒店主体建筑室外绿地种植时，就不宜栽植高大的乔木，以防止其对生态酒店造成过分荫蔽；因为规划绿地强调可进入性，因此在树种选择时，尽量避免枝叶有刺以及过度粗糙，可能对人身体产生伤害的树种；此外，为防止儿童摘尝树木果实而受到伤害，还应避免选择果实有毒或食用后对人体产生不良反应的树种。重点考虑用乡土树种以及野生植物资源，通过合理搭配来营造内涵丰富的生态园林景观。

文化生态涉及的是建筑与地方文化的关系，即对地方历史文化的尊重。酒店所处的地域文化环境使其成为与地域文化环境相结合的独一无二的建筑，而这种独特性正是现代酒店的一个重要的发展方向。

水景是生态造景最重要的内容之一。不论哪一种类型的生态环境，水是最富有生气的因素，无水不活。我国古典园林素有“无水不成园”的说法，水是园林的血液。在生态酒店环境设计中，引入一定面积的水体为主景，这样，使整体环境的设计因为有了水的存在而变得鲜活，有了生气。使原来较为狭小而且各自相对独立的内外空间不仅变的开阔且产生了互动的必然因素；在心理上，给人以扩大与关联的心理暗示。同时，内外环境有了水元素的加入，更有利于小环境的生态平衡，利于小环境中各个生物种群的“共生”。

水景也可用来隔开不同的行动空间，例如在酒店门口设置水景可以有效减低外界的喧闹和繁杂，让人在还未踏进酒店时就能有一种放松的感觉，感受到绿色生态酒店的不同。

水体流速的快慢也要根据周围不同的环境做出选择，比如在饭桌旁的水体就不宜出现水花四溅、水声过大的现象；在安静的空旷区域，选择流速大的动水，会给人带来一种气势磅礴、心旷神怡的感觉。

而且，水体的引入在这里还有其实际用途：水体可以处理、消化掉园中的部分积水，兼做蓄水池之用，以缓解园中排水难的压力。

在绿色生态景观建造中，山石一般分为抽象化的山石景观，如仿日本“枯山水”的山石景观；还有仿自然状态景观，如在酒店内用钢筋混凝土做的假山。山石的作用常具体表现在对比与调和、韵律、主从与重点、联系与分隔等方面。

若酒店选址是在山林间的，天然的山石会成为酒店设计的利用资源。生态绿色酒店的顾客对象往往是外来旅客，因此在设计上更倾向选择本土特色。在山石的材质选择上，要考虑当地的地理风貌，能够表达当地的自然风格。此外山石的设计还需要设计者把握好山石的尺度和酒店室内空间的比例关系，要配合生态绿色酒店的整体轮廓，避免突兀，让人在空间上感受到舒适、自然的味道。

面向未来发展的足够弹性

绿色生态的概念是一种动态的思想，体现在绿色生态酒店中，就是使绿色生态酒店面向未来发展时具有足够的弹性。首先，楼体应具有可生长性，包括基础的预留量、楼地板对承重的预考虑、周边环境的生长预留地等；其次，预留相应的管道空间，包括水、电、通讯的发展空间，在进行设备竖井、机房、面积、层高、荷载等设计时留有发展余地；第三，使用可更新的建筑内外饰面构造方式，使用耐久性强的建筑材料，同时考虑室内装潢的可变化性，便于对建筑保养、修缮、更新设计。

当然建筑理论与实践是有差距的，原则仅仅是一种理想，如何找到与之相适应的技术支持，才是绿色生态酒店向前发展的动力。

The word "green" advocates a new fashion and leads a new lifestyle. Green elements in hotel industry can be found everywhere. With the same theme – "green", those green elements are presented in different forms. Compare to other elements involved in hotel brand, green and ecological design for hotel is an important way to develop an international brand.

Green Design for Hotel Brand

Nowadays, hotels have never been the simple inns or stations during journey. Under great social and living pressures, modern people have borne strong desire for an eco and comfortable environment. And people who live in the depressive concrete buildings will always long for beautiful natural landscape. Thus, green and ecological hotels have become the new trend in hotel industry, and their new contents and typical styles are developed accordingly. Today, "green and ecological hotel" represents a new lifestyle around the world.

Green and ecological hotels refer to the ones designed in ecological principles. The building spaces, shapes and colors are well designed to combine with the surroundings, providing a green, ecological and comfortable place for entertainment, business banquet and tourism. Green and ecological hotels have satisfied people with beautiful landscapes, ecological ideas and wholesome environment, bringing them close to nature.

Regional Features: Essential Elements to Achieve Green and Ecology

The design idea for green and ecological hotels emphasizes minimum destroy to the eco system from the very beginning of a project – site selection. It makes energy saving, environment protection and neighborhood development part of the hotel business. With the eco design requirements, regional features become the essential elements to realize a green and ecological hotel. According to this idea, it should pay attention to ecological development, follows the topography of the site and takes advantages of the surrounding environment; the architectural design should consider the local conditions and local culture, trying to use local materials for the easy restoration of an eco environment. Therefore, the consideration of regional features shows a kind of eco idea. Hapuku Lodge in the famous Kaikoura of the east New Zealand has carried out this idea. At Hapuku Lodge and Tree Houses, it takes full advantages of the local forest, using trees as the roofs to keep harmonious with the nature. It shares the travelers appreciation for the environment. And another hotel Jean-Michel Cousteau's Resort, which is also inspired by this idea, has taken advantage of the tropical sun and built the grass houses. It's also taken advantage of the ideal location to develop underwater exploration and boast a marine biologist, achieving the harmony between human beings and nature。

The consideration of regional features shows a kind of eco idea. It pays attention to ecological development, follows the topography of the site and takes advantages of the surrounding environment; the architectural design should consider the local conditions and local culture, trying to use local materials for the easy restoration of an eco environment.

Cultural and Ecological Connection Between Architecture and Its Surrououndings

The generalized ecological concept includes not only the natural environment, but also artificial environment and historic-cultural environment. The harmony between building and local historic culture is cultural ecology which is also one of the topics to design green & eco-friendly hotel. Cultural ecology involved the relationship between the building and local culture, namely a respect for local history and culture. On one hand, green & eco-friendly hotel not only reflects the local culture, but also keeps a sense of whole with its historical environment and cultural background, so as to further strengthen and protect the historical environment and local context rather than destroy and weaken them; on the other hand, cultural context of the place where the eco-friendly hotel locates endows the hotel some local traditional cultural characteristics, which makes the hotel building the unique combination of geographical and cultural environment, and this unique feature leads an important direction for modern resort hotel. There are two main skills are adopted in the process of architectural design: one is to use traditional architectural form, which not means to imitate the surrounding historic buildings but to make some innovations on the base of traditional form to fit in modern function and structure; the other one is to abstract and simplify historical symbols, and then refine them to create a new form that has the same visual effect as that of the old one, expressing cultural continuity.

The mountain hotel in Ningbo Wulongtan Resort, one of the successful examples, learns from the charm of the traditional Chinese landscape, placing itself in the natural environment in a low-key way, through the

> Cultural ecology involved the relationship between the building and local culture, namely a respect for local history and culture. Cultural context of the place where the eco-friendly hotel locates makes the hotel building the unique combination of geographical and cultural environment, and this unique feature leads an important direction for modern resort hotel.

integration of modern technology and traditional culture to achieve a rich geographical performance skill.

Another example is the Rogner Bad Blumau Resort in southeast Austria, its motto is "subtle differences", however, the visual impact it brings is quite strong. The hotel is like a small town, divided into different apartment blocks, meandering road will lead you to all sorts of unexpected landscape. Rogner Bad Blumau is not only a spa hotel, but a complete artwork, designed by the famous Austrian architect Friedensreich Hundertwasser. Rogner Bad Blumau is a voyage of discovery. Leafy roofs, round shapes, colorful façades and golden domes, surrounded by fields and meadows, create a living work of art.

Ecological Technology Adjusting to Local Conditions

The architectural design of green ecological hotel should always highlight the recreational characteristics. It should keep practical attribute as the principle to avoid being large and all inclusive. The commonly used ecological technologies include natural lighting, ventilation, shading technology, ecological water-saving, energy-saving and emission reduction technology, etc.

Appropriate technology, also called intermediate technology or medium technology, advocates to be adjust to local conditions. The intermediate technology has much higher production efficiency than the soil technology (in the condition of the recession), and much cheaper than the advanced technology with intensive modern industrial capital. From the perspective of regional attribute of green ecological hotel

design, appropriate architectural technology is a technological manner specific to the geography, climate, regional attribute and culture. It inherits and develops the traditional elements and critically absorbs the foreign advanced practical technology. The architects can use local available materials and go on construction by the local traditional handicraft. From the perspective of design level, the architects handle well the relationship of contrast and unity between the modern material and local traditional materials and combine them with modern technique to achieve the unity between the buildings and the surrounding environment, and also meet the cost requirements. The design scheme of Alfla Viuas Uluwatu Hotel in Bali, Indonesia is an example that respects and protects the environment, and its overall planning respects the existing terrain, avoiding earth cutting and filling. The big trees are retained or removed. The local plants cultivated on the project base, rather than exotic plants, are used in the landscape construction. It also promotes the local animals and birds to stay in the local region to maintain the balance of the ecosystem. From the initial design stage to the final products provi–

sion, the environmental behaviors strictly adhere to the requirements of environmental protection and the building materials are all green non-polluting materials.

Creating Conditions for Communication with Natural Environment

Greening ecological hotel as the bridge connecting users and natural environment is expected to lead as much natural elements as possible to the users, which is a key interpretation of ecological principle. Architectures are not the burden and shelter between human and nature any more; it will provide a new lifestyle. Interior space in buildings is full of fresh air, soft and gentle sunlight instead of fluorescent lighting. Natural elements are adopted to create conditions for communication between users and natural environment, which is another target and pursuit of ecological resort hotel. Three key natural elements are required in the ecological environment design: plants, waterscape and rocks.

Plants are absolutely required in the design of greening ecological environment. Various kinds of plants are generally arranged to build different landscapes, with functions of environment beatification, spiritual delight, environmental improvement and air purification. Plants work with landscape to create ecological environment. Given the lighting demand for plant growth inside the hotel, tall trees will not be adopted as the plants outside close to main buildings of ecological hotel, preventing too much shading on hotel. Accessibility is highlighted in the planning green land, trees with sting or being too rough, possibly harmful to human will mot be selected. To prevent children injure from picking and eating fruits, trees with poisonous fruits or having adverse effect after eating shall not be adopted. Native tree species and wild plant resources are adopted as a key consideration to coordinate and build up ecological garden landscape with rich connotation.

Waterscape is one of the most important parts in ecological landscaping. Water is the most vivid and essential factor in any type of ecological environment. Chinese ancient garden are renowned for a saying of "no water, no garden", water is the running blood in a garden. The entire environment becomes vivid and fresh due to the application of water, a certain area of waterscape. As a result, the former narrow internal and external spaces separated from each other become broad

and associated, generating a psychological hint of expansion and association. In addition, water factor in internal and external environment propitious to maintain the ecological balance and symbiosis of biological populations in microenvironment.

Waterscape could be used to separate different event spaces, i.e. waterscape in the hotel entrance could efficiently shield against the external noise and rumpus, offering a sense of relaxation to people even before they enter the hotel, encouraging them to experience the difference of greening ecological hotel.

The speed of running water is variable for different environments surrounding, for instance, water with splash and loud sound is not proper by a table, while quiet open area matching running water with high speed will present majestic and awe-inspiring, carefree and joyous sense.

In addition, the water body is adopted for actual use: it could carry away part of stagnant water and act as a reservoir, releasing the drainage pressure in the garden.

In the construction of greening ecological landscape, rocks are classified as abstract rock landscape, i.e. rock landscape imitated Japanese "dry landscape", and nature-imitation landscape, i.e. rockery made of reinforced concrete. Rocks are generally used to present the contrast and harmony, rhyme, principal and sub-principal, association and separation etc.

Natural rocks will be a great resource of hotel design if it is located in mountain forest. Customers of greening ecological hotel are mostly visitors from outside, thus the hotel design prefers to local characteristic. In the selection of rock materials, the designers have to consider local geographic features, and to show the local natural style. Besides, the rock scale and the ratio of interior space in hotel should be mastered to match the entire outline of the hotel, providing sense of comfort and nature.

Great Flexibility for Future Development

The concept of green ecological idea is a dynamic concept reflected in the green ecological hotel, which is to gain enough flexibility for the future development of the green ecological hotel. Firstly, the building should be adaptable, including the reserved amount of the base, preliminary consideration of floor bearing and reserved land for the surrounding environment development, etc. Secondly, it should obligate some corresponding pipeline space, including water, electricity and communication space for development, and reserved space for the design of shaft equipment, machine room, area, height and load, etc. Thirdly, use construction methods of renewable internal and external building surface, use durable building materials and consider the changeability of the indoor decoration, so that it can be convenient for maintenance, repair and the update of architecture design.

Of course, there is a gap between the architectural theory and practical situation; the principle is just a kind of ideal, how to find the corresponding technical support is the moving power for the green ecological hotel development.

Le Méridien Istanbul Etiler | 伊斯坦布尔艾美酒店

Keywords 关键词

Modern Style 现代风格

Aesthetic Characteristics 美学特征

Decorative Vision 装饰视觉

Leisure Vacation 度假休闲

酒店地址：土耳其伊斯坦布尔艾提雷区 Cengiz Topel Caddesi 大道 39 号

电　话：+902123840000

Address: Cengiz Topel Caddesi No. 39 Etiler Istanbul, Turkey

Tel: +902123840000

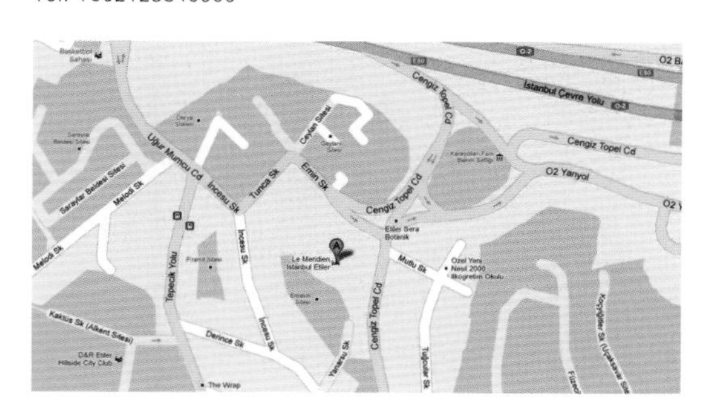

品牌链接

艾美酒店（Le Méridien），又被称为美丽殿酒店，是一家跨国的酒店品牌，原总部设于英国，属于喜达屋酒店及度假酒店国际集团。艾美酒店在全球 50 多个国家中有超过 120 个分店，主要设于欧洲、非洲、中亚、亚太地区和美国等地的热门旅游景点附近。

1972 年，法国航空建立了 Le Méridien 这个酒店品牌，以"提供客户宾至如归的感受"为目标。第一家艾美酒店（Le Méridien Etoile）于巴黎开幕，共有 1000 个房间，2 年内，共有 10 个分店纷纷于欧洲和非洲开幕。美丽殿酒店营运的前六年，开幕的酒店数目提高到 21 家，除了欧洲和非洲外，据点也拓展到中美洲、加拿大、南美洲、中亚等地。到了 1991 年中，艾美酒店的总数提升到 58 家。1994 年底，艾美酒店被英国的福特集团（Forte Group）买下，1996 年又被 Granada Group plc 收购，而福特集团的母公司格拉纳达集团（Granada Group plc）和康柏司集团（Compass Group plc）于 2000 年夏天合并，又在 2001 年 2 月分家，福特集团原有的三家酒店品牌，最后全归 Compass Group plc 所有。2001 年 5 月，野村财阀宣布以 19 亿英磅买下，并于 2001 年 2 月将艾美酒店与 Principal 酒店合并。2003 年 12 月，雷曼兄弟控股公司收购了艾美酒店。2005 年 11 月 24 日，艾美酒店和相关企业被喜达屋酒店及度假酒店国际集团并购。

About Le Méridien

Le Méridien Hotel is an international hotel brand and its original headquarters is in the UK, which belongs to the Starwood Hotels & Resorts Worldwide, Inc. Le Méridien Munich Hotel owns more than 120 branches in more than 50 countries around the world, mainly located near the popular tourist attractions in Europe, Africa, Central Asia, the Asia-Pacific region and the United States.

The Le Méridien brand was established in 1972 by Air France "to provide on a home away from home for its customers." The first Le Méridien property was a 1,000-room hotel in Paris — Le Méridien Etoile. Within two years of operation, the group had 10 hotels in Europe and Africa. Within the first six years, the number of hotels had risen to 21 hotels in Europe, Africa, the French West Indies, Canada, South America, the Middle East and Mauritius. By 1991, the total number of Le Méridien properties had risen to 58. In late 1994, Le Méridien was acquired by UK hotel company Forte Group, which in turn was acquired by Granada plc in 1996. Through a merger in the summer of 2,000 between Granada Group and global contract caterer Compass Group— and the subsequent de-merger of the two companies in February 2001— the ownership of the Forte Hotels division and its three brands (Le Méridien, Heritage Hotels and Posthouse Forte) passed solely to Compass. In May 2001, Nomura Group announced the acquisition of Le Méridien Hotels & Resorts from Compass Group plc for £1.9 billion and Le Méridien was merged with Principal Hotels, which was acquired in February 2001. In December 2003, Lehman Brothers Holdings acquired the senior debt of Le Méridien. On November 24,2005, the Le Méridien brand and management fee business was acquired by Starwood Hotels & Resorts. The leased and owned real estate assets were acquired in a separate deal by a joint venture formed by Lehman Brothers and Starwood Capital. In 2011, Le Méridien opened up its 100th hotel in Coimbatore, India. In mid 2012, during the search conducted by Maradu Municipality, the food safety officials seized damaged food from star hotels including Le Méridien Kochi.

MERIDIEN

Le MERIDIEN

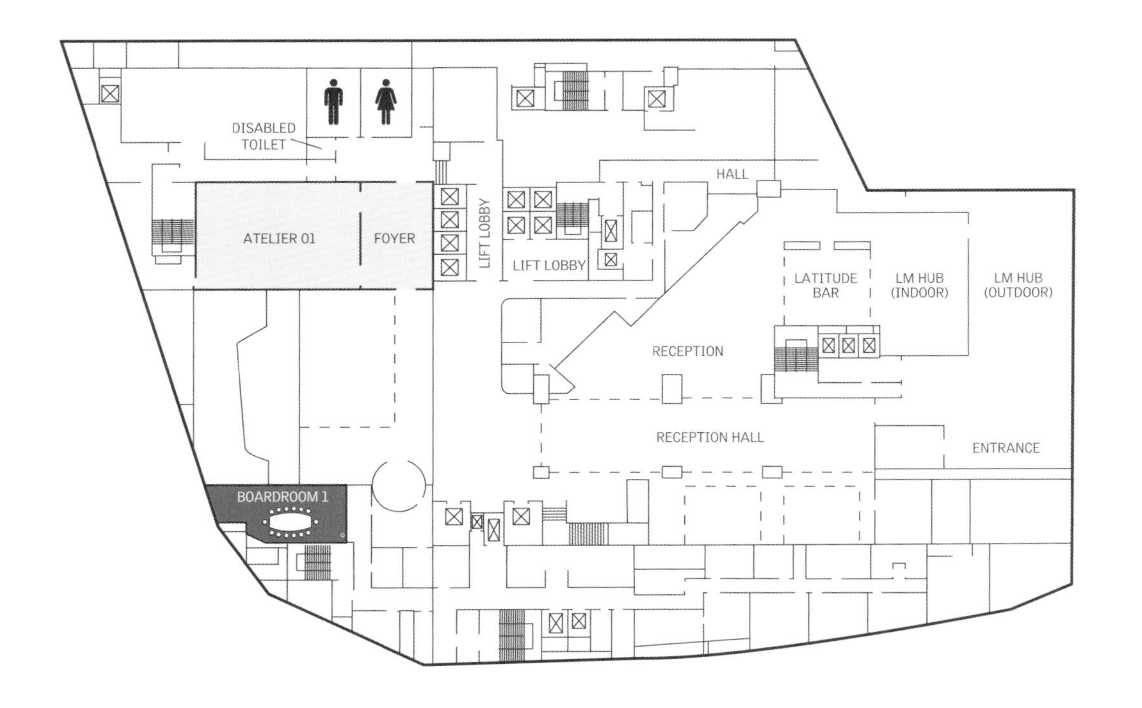

酒店外观

酒店外观有点写字楼的味道，造型及色调非常现代，整体颜色采用冷色调，而不是运用艾美经典的红色调，淡雅的灰色调给人以舒心的感觉。家具中规中矩中带些时尚感，建筑力求散发不一样的视觉效果，不论软装、硬装、灯光、材质、色彩都搭配得协调，也颇具特色。

Exterior

Very modern facade with a little office taste, shape and tone, the overall color of cool colors, rather than the use of Le Meridien's classic red tones, elegant gray tone gives a comfortable feeling. Furniture quite satisfactory with some fashion, the architecture strives to distribute different visual effects, the soft or hardware, lighting, material, color coordination are distinctive.

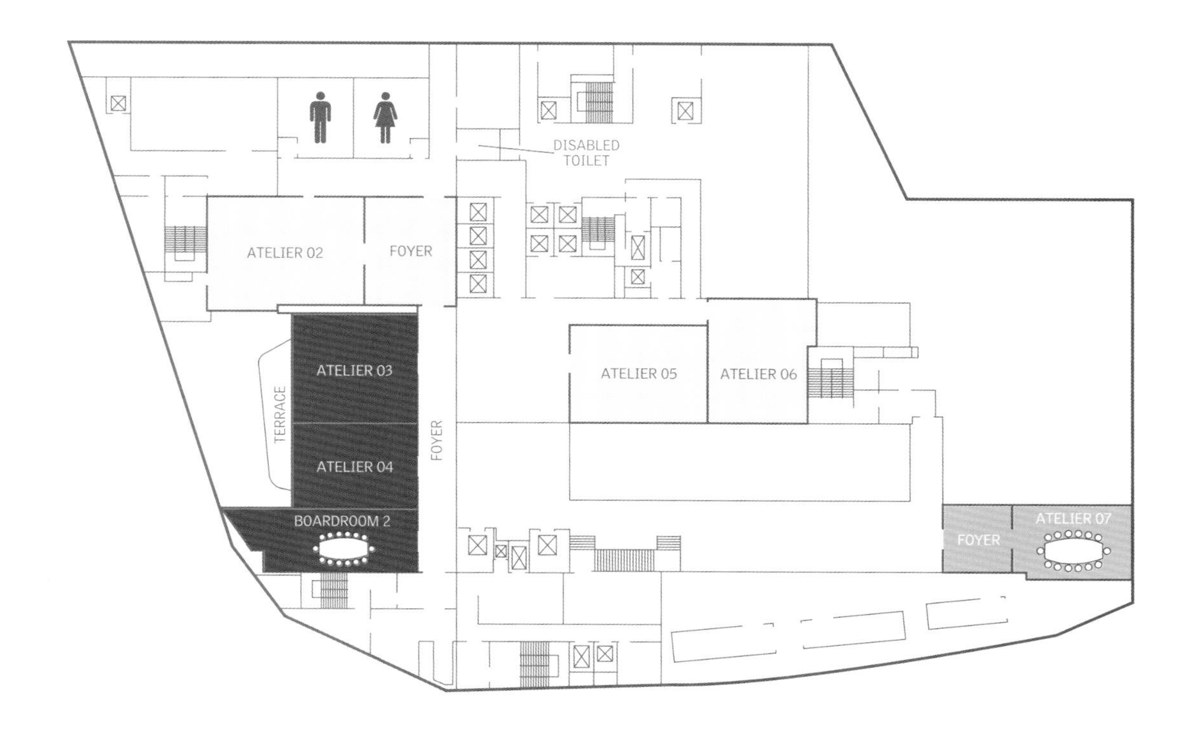
DISABLED TOILET
ATELIER 02
FOYER
ATELIER 03
TERRACE
ATELIER 04
FOYER
BOARDROOM 2
ATELIER 05
ATELIER 06
FOYER
ATELIER 07

酒店概况

伊斯坦布尔艾美酒店由世界著名的室内设计师 Sinan Kafadar 设计，有着标志性的当代建筑风格。它结合了创意设计优势，带给旅客一种独特的体验，反映伊斯坦布尔的启发性和多元文化的氛围。

酒店共有 6 个餐厅和酒吧，1 个健身中心和 1 个水疗中心，3 个室外游泳池，并提供一系列团体活动和水上运动。

Overview

The hotels with iconic contemporary architecture style designed by globally renowned interior designer Sinan Kafadar, combined with Le Méridien's creative design edge, delivers a unique experience, reflecting Istanbul's inspiring and culturally diverse atmosphere. It offers six restaurants and bars, a fitness centre and spa, plus three outdoor swimming pools and a range of group activities and water sports.

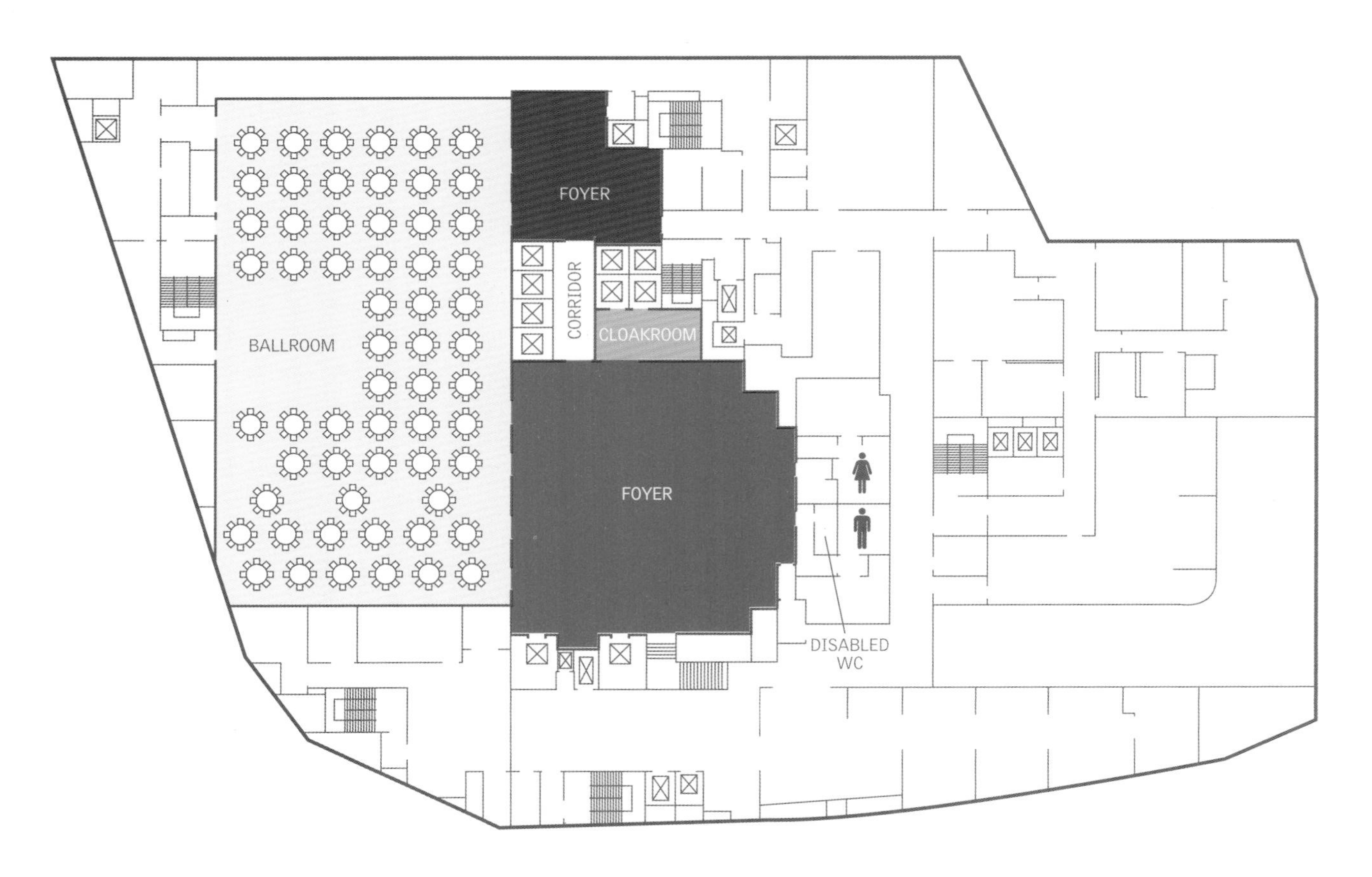
FOYER
CORRIDOR
CLOAKROOM
BALLROOM
FOYER
DISABLED
WC

酒店特色

艾美酒店秉承欧洲优雅传统，融合当代文化，采用现代风格与度假形式相结合，构建出富含精致人文气息、深邃内涵和激动人心的氛围，使每个身处其中的人都能尽享每一刻美好时光，于平淡无奇的日常生活之中发掘美妙事物。

Feature

Adhering to the traditional European elegance and contemporary culture, the hotel combines modern style with holiday form to build a rich exquisite cultural atmosphere and a deeply connotation and exciting atmosphere in which all people can enjoy the moment, explore the wonderful things in prosaic everyday lives.

酒店配套

■ 餐饮

酒店设有6间餐厅和酒吧，在这片绵延6 km的美丽海滩提供餐饮服务。

Services and Amenities

■ Dining

Food + beverage service is offered at this beautiful 6-kilometer beach, shared with the rare visitor strolling from a nearby resort.

■ 休闲活动

酒店设有 1 个健身中心和水疗中心、3 座室外泳池，并提供各种趣味纷呈的团体活动和水上运动。客人还可在清澈透明的海水中静静感受漂浮体验，或通过多种海滩活动尽享无限欢乐，其中包括免费清晨和晚间慢跑、沙滩排球、沙滩足球、沙滩滚球游戏、沙滩高尔夫切杆以及清晨劲走、沙滩伸展运动和独木舟运动。

■ Sports

Guests can quietly float in the crystal clear waters or enjoy multiple beach activities, including complimentary morning and evening jogs, beach volleyball, beach football, beach pétanque, and beach golf chipping, as well as morning power walks, beach stretching, and outrigger canoeing.

■ 海景泳池

酒店的海景泳池是一个风景如画的迷人场所，在此可尽情畅游、锻炼身体，也可参与水上排球、篮球或水球等各种热闹欢腾的趣味游戏。泳池区的多座按摩池，可放松休憩。宁谧美丽的礁湖泳池是在礁湖阴影或明媚阳光下享受悠闲下午漂浮的绝佳地点，此外，还可在此细细品尝清凉健康的美味饮品。企鹅儿童泳池则是孩子们的欢乐王国。

■ Horizon Pool

Horizon Pool provides a picturesque attraction that will inspire customers to swim a few laps, exercise rigorously, or engage in a rollicking game of water volleyball or basketball or even water polo. When finished, they could rest in one of the pool area's multiple Jacuzzis. Steal away to the Lagoon Pool for a lazy afternoon floating in and out of the lagoons' shade or sipping an enticing and healthy drink. Small children reign in our Splash Pool, where they can engage in carefree enjoyment while you watch. A splash pool designed specifically for children to play in is located next to the Penguin Club for children.

THE GLASS HOUSE

■ 客房

酒店共提供11种不同客房：豪华客房、博斯普鲁斯海峡客房、创意客房、行政客房、普通套房、复式套房、行政套房、住宅套房、总统套房、住宅公寓2+1和住宅公寓3+1。

■ Rooms & Suits

There are 11 room categories available: Deluxe Room, Bosphorus Rooms, Creative Room, Executive Room, Junior Suite, Duplex Suite, Executive Suite, Residential Suite, Presidential Suite, Residential Apartment 2+1 and Residential Apartment 3+1.

The Westin Xiamen

厦门威斯汀酒店

Keywords 关键词

Unique Experience 独特体验

Heavenly Bed 天梦之床

Heavenly Spa 天梦水疗

Ideal Location 位置优越

品牌链接

威斯汀酒店及度假村是喜达屋酒店与度假村国际集团的品牌之一，在全球超过 36 个国家和地区拥有 180 多家酒店和度假村。威斯汀酒店及度假村可使客人尽享康乐元素。清爽氛围、创新计划和周到设施有助于为客人提供一次乘兴而来、满意而归的绝佳住宿。现代化的设计、悉心周到的服务和令人活力焕发的氛围成为全球 180 多家威斯汀酒店及度假村的招牌。无论在西班牙享受高尔夫挥杆之乐，在巴厘岛体验惊险浮潜之游，还是在时代广场悠闲观光游览，威斯汀总能带给顾客与众不同的完美体验。

About Westin Hotels and Resorts

With more than 180 hotels and resorts opened in 36 countries and areas, Westin Hotels and Resorts is a hotel brand of the Starwood Hotels and Resorts Worldwide Inc., which is the most high-end hotel company in the world. The guests indulge in elements of well-being in Westin. The refreshing ambiance, innovative programs and thoughtful amenities can provide the guests with a stay that leaves one feeling better than when arrived. More than 180 hotels and resorts worldwide are defined by modern design, instinctive service and rejuvenating atmosphere. Whether golfing in Spain, snorkeling in Bali, or sightseeing in Times Square, Westin delivers a perfect experience unlike any other.

酒店地址：中国福建厦门思明区仙岳路 398 号
电　话：86-592-337 8888

Address: No. 398 Xianyue Road, Siming District, Xiamen City, Fujian Province, China
Tel: 86-592-337 8888

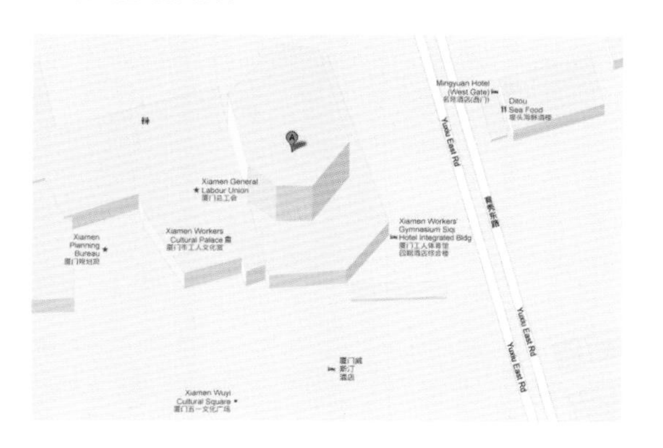

WESTIN

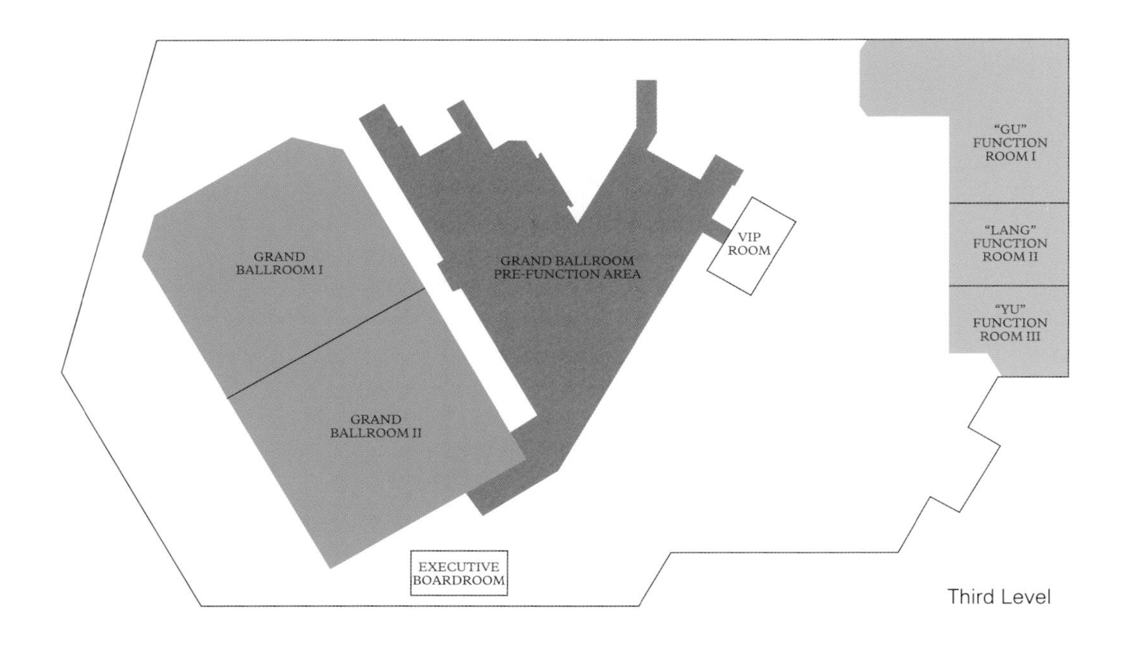

酒店开发管理

厦门威斯汀酒店由厦门翔业集团旗下的佰翔集团投资打造。厦门正在显露出成为中国最大奢侈品消费市场之一的潜力，这里的人们也在探寻不一样的生活方式。而威斯汀品牌完全符合这个市场需求，它在全中国都有着强劲的品牌号召力，通过独特的酒店文化和品牌项目，厦门威斯汀酒店也一定能为造访厦门的客人提供活力复苏的旅行体验。厦门威斯汀酒店作为2012年度中国区第二家开幕的威斯汀酒店，进一步扩大了威斯汀品牌在中国以及全球的业务版图。

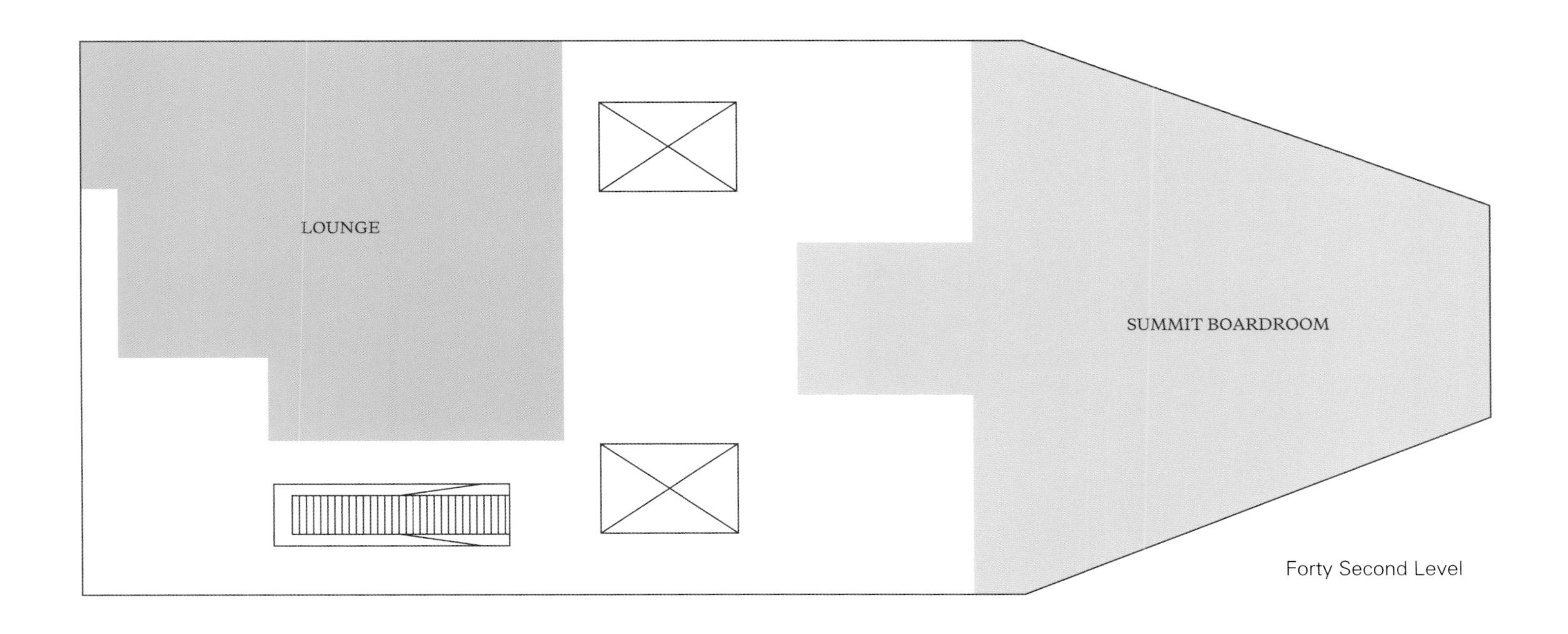

Forty Second Level

Development and Management

The Westin Xiamen is invested by Xiamen Fliport Hotel Group and operated by the Westin Hotels and Resorts. As Xiamen shows its potential to be one of the biggest luxury consumption market in China, people here are trying to find different lifestyles. Westin just well fits the market with unique hotel culture and its brand reputation. It will provide guests with a refreshing travel experience. The Westin Xiamen is the second Westin Hotel that opened in 2012 in China, which further expands the Westin brand in China as well as in the world.

酒 店 概 况

酒店地处商业中心，设有宽敞与高端的宴会及会议场所，提供国际水准的美食佳肴，更有标志性的威斯汀产品使客人焕发身心活力。厦门威斯汀酒店距离厦门高崎机场约15分钟车程，城市商业区和五一文化广场等各种当地景点近在咫尺，这是一座融艺术、运动、购物和娱乐于一体的时尚建筑群体。

Overview

The Westin Xiamen is strategically located in the central business district, offering guests with high-end facilities for banquet and conference, delicious food of international standard as well as the typical Westin products to make one re-freshing. The Westin Xiamen is just 15 minutes by car to the Gaoqi International Airport, and the city's business areas as well as local attractions, such as the Wu Yi Plaza.It's an ultra-modern complex that combines arts, sports, retail, and entertainment under one roof.

酒店特色

威斯汀酒店及度假村提供提升宾客健康的服务和设施，让宾客在离店时体验焕然一新的美妙感受。这座高耸入云的地标建筑为欣赏风景秀丽的筼筜湖和郁郁葱葱的仙岳山提供绝佳的观景视角，厦门城市美景在此尽收眼底。

Feature

Westin Hotels and Resorts will provide the services and facilities to promote guest's health that leaves one feeling better than when arrived. Towering above the city, the Westin Xiamen offers spectacular views of the scenic Yu-andang Lake, the lush Xianyue mountains as well as the bustling and vibrant cityscape of Xiamen.

酒店外观和室内

酒店楼高45层，拥有304间豪华客房和套房，从酒店内既可远眺厦门筼筜湖景及繁茂山景，亦可一览城市景观。从客人抵达酒店开始就能体会到各式感官迎宾，如酒店简洁亮丽的装潢、舒适放松的音乐、飘逸怡人的白茶香氛、大堂的鲜明灯光等。南北两个大堂分别能为非团队客人和团队客人提供简便高效的入住流程。

Exterior and Interior

The hotel tower is 45-storey high with 304 luxury rooms and suites, providing spectacular views of the scenic Yuandang Lake, the lush Xianyue mountains as well as the bustling and vibrant cityscape of Xiamen. When entering into the hotel, guests will receive warm welcome from every corner, such as the elegant decoration, relaxing music, nice fragrance of white tea, bright lights in the halls and so on. The south hall and north hall are set separately for the independent guests and team guests, providing high-efficiency check-in and check-out service.

酒店配套

■ 餐饮

厦门威斯汀酒店的 1~4 层为公共区域，专门为客人提供不同特色的国际美食和丰富的娱乐世界。中国元素餐厅供应耳目一新的粤菜和用传统方式烹饪的中式佳肴，知味全日餐厅供应营养丰富的创意美食自助餐，Qba 以其音乐、美食以及种类繁多的葡萄酒和鸡尾酒让顾客尽享拉丁文化，大堂吧的鸡尾酒或下现煮咖啡则可让顾客彻底放松身心。厦门威斯汀酒店提供 24 小时的客房用餐服务，并且向客人供应补充其身体营养的活力食品。

■ 会议

酒店会议面积超过 1 500 m²，可满足不同类型的宴会及高端会议的需求。无柱大宴会厅可灵活分隔成两个独立空间，为客人提供理想的宴会场所。另外 10 间独立会议室配备了各种会议专用设施，同时酒店会议楼层还设有高速无线上网和专业会议服务前台，随时满足客人在会议上的需求。

Service and Amenities

■ Dining

The 1st to 4th floors of the hotel are dedicated as the public area, with the dining and entertainment options for guests to explore. Seasonal Tastes is an all-day-dining restaurant presenting nourishing and imaginative menu creations. Five-Zen features new-look Cantonese fare and a range of Chinese dishes prepared using authentic methods. Qba lets guests immerse themselves in Latin American culture with vibrant music, food and an extensive wine and cocktail list, and Lobby Lounge is just the place where guests can relax and elevate their senses with an end-of-day cocktail or just a freshly brewed coffee. In room dining is available 24 hours. The Westin Xiamen is proud to offer SuperFoodRx™ among its culinary offerings; proven to nourish and replenish the body.

■ Meetings and Events

Towering above the central business district, the Westin Xiamen offers more than 1,500 square meters of meeting venues for events of all shapes and sizes. The contemporary spaces can be transformed as required a lavish gala or a board meeting. The dedicated team of professionals will make sure customers have what they need for successful events. Keep minds sharp and attention focused with a menu of energizing and tasty dishes. The banquet services team will oversee every detail, ensuring the guests are kept refreshed throughout their events.

■ 特色服务

厦门威斯汀酒店除了提供威斯汀酒店品牌的标志性服务和产品，如威斯汀天梦之床（Heavenly® Bed），威斯汀健身（Westin WORKOUTTM），天梦婴儿床（Heavenly® Crib®）外，更设有厦门独一无二的威斯汀天梦水疗 (Heavenly Spa by WestinTM)，融合古代中式疗法和现代水疗概念，在水疗中心高雅恬静的环境当中，为客人提供一系列消除疲劳，恢复身心平衡的疗程。

每三个楼层拥有一个共同的中庭，其温馨的氛围可以让客人舒缓身心与轻松社交。行政酒廊位于 40~42 层，除了可以从高空俯瞰厦门城景，这里还提供个性化的服务，如快速办理登机牌，擦鞋服务，Wii 与 XBOX 体感游戏机，高速网络接入，豪华早餐，鸡尾酒时刻与更多生动体验。42 楼的董事会议室则是商务会谈的绝佳私密场所。

■ Special Offer

Find peace in a busy traveling schedule by using the Westin Xiamen's unique Flight@Westin service. Guests can enjoy express check-in and boarding pass issue at the hotel lobby counter or Executive Club lounge. Real-time flight updates on guest room television and hotel lobby allow guests to plan for a hassle-free journey.

The hotel also takes care of the guests with other services like Concierge, Travel Desk, Business Center, Gift Shop and Valet Parking. Careful attention is paid to the most important elements of their stay to make sure the guests feel better than when they arrived.

■ 客房

酒店304间客房与套房均配备高雅装潢及先进设施，包括高速互联网接入，42寸液晶纯平电视，专用的iPod数码埠确保客人的数码装备始终保持连接。宽敞的办公区域配上大办公桌、Herman Miler人体力学座椅和咖啡机，让商务客人在工作中事半功倍。豪华的大理石浴室以玻璃门分隔出不同区域，包括装有热带雨林花洒的淋浴间和配备Laufen品牌的浴缸区域。其他威斯汀特有天梦用品包括威斯汀白茶浴盐、芦荟乳液、舒适浴袍和拖鞋。

■ Rooms

Arrival has never felt this good at the Westin Xiamen, with its unique architecture of 45 stories and a total height of 169.3 meters. Each of the 304 inspiring guest rooms and suites are designed to let the guests be at their best. Get an amazing night's rest in our Heavenly Bed™. Other amenities include high-speed Internet access, a generous work area with an ergonomic chair and 42-inch flat screen LCD television - everything that is needed for a rejuvenating stay.

Palace Hotel Tokyo

东京皇宫酒店

Keywords 关键词

Modern Style 现代风格

Luxurious and Comfortable 豪华舒适

Large Space 大空间

Natural Lighting 自然采光

品牌链接

东京皇宫酒店隶属于皇宫酒店有限公司，是一家成立于 1964 年的私人股份制集团。该公司的创始人和第一任管理人是政友吉原。该酒店还荣获过 1963 年的建筑业协会奖，源于其将现代建筑风格与日本美学成功地融合在一起。

About Palace Hotel

Palace Hotel Tokyo is owned by Palace Hotel Co., Ltd., a private consortium of shareholders first formed in 1961. The company's founder and first president was Masatomo Yoshihara. The Palace Hotel was awarded the Architectural Industry Association Prize in 1963 for its success in blending modern architectural style with Japanese aesthetics.

酒店地址：日本东京千代田区丸之内 1-1-1 号
电　　话：+81 (0) 3 3211 5211

Address: 1-1-1 Marunouchi, Chiyoda-ku, Tokyo, Japan
Tel: +81 (0) 3 3211 5211

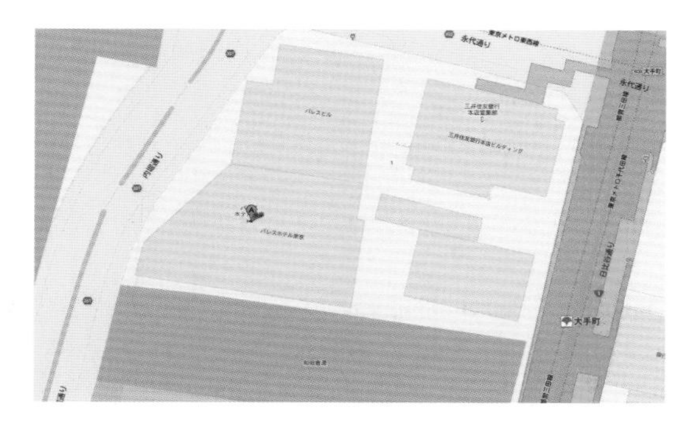

酒店开发管理

皇宫酒店于2012年开业，由东京一些城市最知名的房地产商所掌管，作为城市最具标志性的酒店，可以追溯到半个世纪前。如今的东京皇宫酒店是在原先的Teito酒店和皇宫酒店的旧址上建立起来的。2009年，在为世界各地的旅客服务了近半个世纪后，皇宫酒店歇业重新装修，准备更名为东京皇宫酒店，使其成为城市里最新的一道豪华的风景线，并在承诺继续保持其独立性和提供最好的本土酒店服务的基础上，发扬其殷勤好客的品质。

Development and Management

It was opened in May 2012, the 290-room Palace Hotel Tokyo is heir to some of the city's most exclusive real estate, with a legacy going back more than half a century as one of the city's most iconic hotels. The contemporary reinvention of Palace Hotel Tokyo resides on the very same site on which its two predecessors – Hotel Teito and Palace Hotel – were situated. In 2009, after having serviced travelers for nearly half a century, Palace Hotel closed its doors in preparation for its reincarnation as Palace Hotel Tokyo – the newest addition to the city's luxury scene and set to redefine hospitality with its continued commitment to preserving its independence and delivering the finest home-grown hospitality.

酒店概况

酒店位于历史悠久的江户城堡前，毗邻日本皇居和护城河，地处该城市的金融和商业区中心地带。近在咫尺的大手町站、东京中央车站和著名的银座购物区，使旅客更容易发现东京皇宫酒店的所在，也提供了更舒适的居住环境。

Overview

Situated in front of the expansive grounds of the historic Edo Castle, next to the Imperial Palace and moats, the hotel is at the heart of the financial and business district of the city. Its close proximity to Otemachi Station, Tokyo Central Station, and the famous Ginza shopping district, makes the Palace Hotel Tokyo the most accessible and ideal hotel for discerning travelers.

酒 店 特 色

酒店拥有 8 个宴会厅，有时可以拆分成 15 个，位于第二层和第四层，总建筑面积 2 596 m^2。所有客房提供自然采光措施，往外可以看到郁郁葱葱的园林景观。酒店内都配备了无线互联网，还有 4~7 m 高的天花板。“葵”号房是最大的宴会厅，可容纳 1 200 个客人。其玻璃墙正对着 Wadakura 护城河，使得宴会厅空间看起来更开阔。

Feature

Eight banquet rooms, 15 when divided, are located on the 2nd and 4th floors and have a total floor space of 2,596 m^2. All rooms offer natural lighting, afford beautiful views of the lush green landscape, are equipped with Wi-Fi Internet, and feature 4 to 7 meters high ceilings. “Aoi”, the largest ballroom accommodates up to 1,200 guests for a reception and its glass wall facing the Wadakura Moat makes it feel like a wide-open space.

酒店室内

酒店拥有 278 间客房和 12 间套房，从房间宽大的玻璃窗往外看，客人可以尽情享受窗外宁静优美的松树园林和皇宫周围潺潺的护城河水。东京皇宫酒店有 12 种不同类别的豪华客房，面积在 45 m^2 到 55 m^2 不等，套房则是 75 m^2 到 255 m^2 不等。大多数房间都配备有开放式浴室，有单独的泡浴和淋浴功能，超过一半的客房和套房有开放的楼梯和阳台。从每一个客房里面都可以欣赏到东京皇宫花园的美景，并能俯瞰周围建筑的空中轮廓线。

Interior

278 guestrooms and 12 suites offer picturesque views of the Imperial Palace gardens and moats and the surrounding Tokyo skyline beyond. The 12 categories of accom-modations include guestrooms ranging in size from an ample 45 square meters to a spacious 255 square meters. Most of the hotel's rooms have open-style bathrooms with separate soaking tubs and showers, and more than half feature open terraces and balconies - a true rarity in Tokyo.

酒店配套

■ 餐饮

酒店全天候开放的西餐厅、酒吧和休闲区提供了东京最具多样性的饮食选择。每个场地是一个单独设计的空间，具有各自不同的情调，如皇家宫殿的大厅酒吧和给人欢乐的休闲区。

■ 庆典及会议

酒店的位置得天独厚，正好处于市中心的繁华地带，使其成为大型展览活动、多日会议、亲人朋友聚会和社交活动等等的首选之地。在这举行婚礼也是一件让人最难忘的事情。

Services and Amenities

■ Dining

The all-day dining restaurant Grand Kitchen and the hotel's wonderfully situated bars and lounges round out Tokyo's most dynamic eating and drinking collective. Each venue is an individually designed space that cultivates a mood of its own, from the brooding refuge of Royal Bar to the cheerful disposition of The Palace Lounge.

■ Celebrations and Other Events

Palace Hotel Tokyo's enviable address at 1-1-1 Marunouchi right in the heart of the city has made it a highly sought-after destination for hosting everything from large exhibitions to multi-day conferences, from intimate gatherings to spectacular social events. Weddings are always the most memorable of affairs.

■ 水疗中心

水疗中心位于酒店的第五层，占地1 200 m^2。其中，东京依云水疗中心的五个治疗室、一个水疗套房、单独的男女休闲区，组成了该城市最活跃的温泉中心。法国先进技术和亚洲疗法给东京提供了一个最精致的水疗体验舞台。

■ 运动健身

酒店的第五层设有室内游泳池，配有按摩浴缸和一个127 m^2 的健身房。健身房配备了泰诺品牌的健身设备和运动系统的设备。

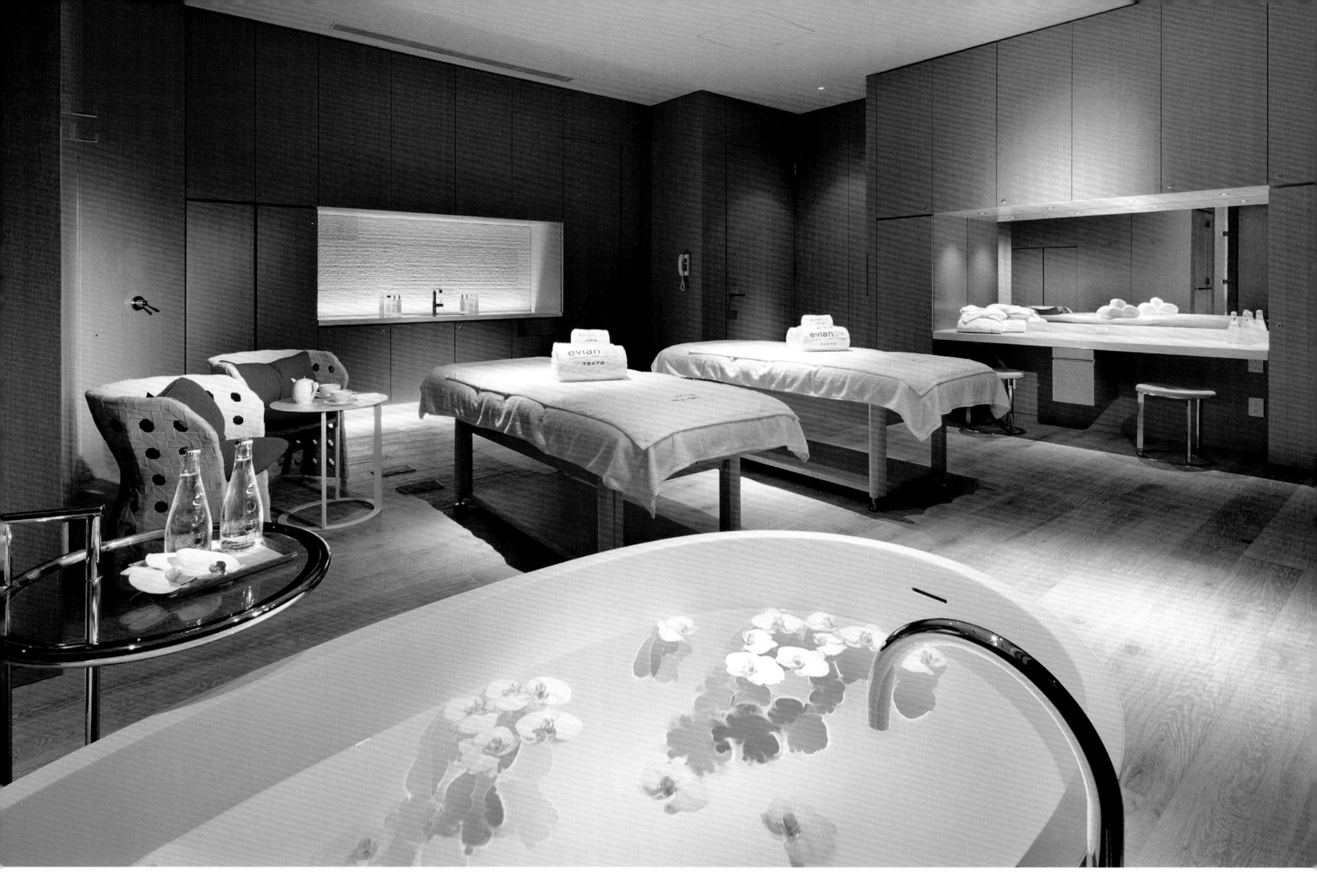

■ Spa

Occupying 1,200 square meters of space on the fifth floor, evian SPA TOKYO's five treatment rooms, one spa suite and separate men's and women's relaxation lounge form the heart of the city's most ambitious new spa. French savoir-faire and Asian therapies set the stage for one of Tokyo's most refined spa experiences.

■ Fitness Center

The hotel's fifth floor is also home to an indoor swimming pool with Jacuzzi and outdoor terrace as well as a 127-square meter Fitness Room outfitted with equipment by Technogym and Life Fitness and a Kinesis System.

Eastern Mangroves Hotel & Spa | 东方红树林安纳塔拉水疗酒店

Keywords 关键词
Luxury 奢华
Spa 特色水疗
Arabic Culture 阿拉伯文化

品牌链接

安纳塔拉，梵语意为“无穷无尽”，象征着自由、运动与和谐，正是“安纳塔拉体验”的核心所在。每一座安纳塔拉都是五星级度假村，从所在地丰富的文化传统、历史古迹与自然美景中汲取精华。因此，每种体验都是一场独特的探索与灵感之旅，并且是安纳塔拉所独有的。

About Eastern Mangroves Hotel & Spa

Anantara, a word in Sanskrit means “without end” and evokes the freedom, movement and harmony that are the spirit of the Anantara Experience. Each Anantara Resort draws its strength from the rich cultural traditions, historic heritage and natural beauty of its destination. As such, every experience is a unique voyage of discovery and inspiration that is distinctly Anantara.

酒店地址：阿联酋首都阿布扎比萨拉街东方路
电　　话：+971 2 656 1044

Address: Eastern Road, Salam Street, Abu Dhabi, United Arab Emirates
Tel: +971 2 656 1044

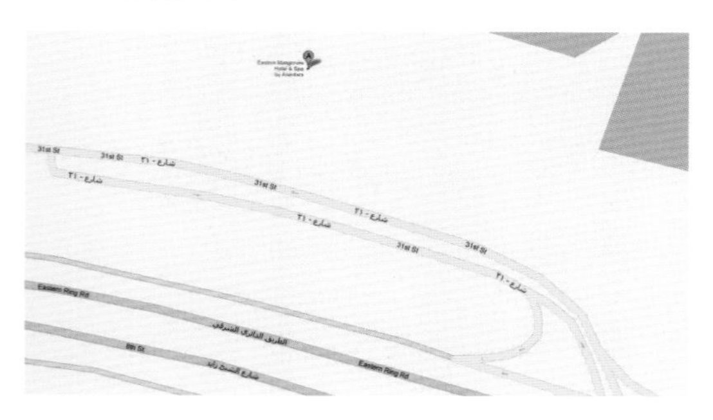

القرم الشرقي
EASTERN MANGROVES

品牌发展历程

2001 年，第一座安纳塔拉度假村在泰国历史悠久的海滨度假胜地华欣建成。独具特色的豪华水疗度假村让客人置身于传统的泰国村落，更加贴近泰国丰富的文化与历史传承，更能感受当地的风土人情；并且通过烹饪课程、水果雕刻展示、每周的水上市场甚至传统武术泰拳教授，以互动方式加强这种体验，成为水疗度假村的经典。

从那时起，“安纳塔拉体验”被带到了泰国最北部的金三角地区，推出了独一无二、世界著名的“安纳塔拉大象营体验”；还有苏梅岛，让人深深感受到泰国南部的丰富文化。2006 年，随着马尔代夫安纳塔拉的建成，“安纳塔拉体验”走向了全球；客人有机会体验真正的荒岛逃逸之旅，五星级的度假村让人犹如置身梦境。

此次得以成功揭幕，并受到来自全世界客人的好评，安纳塔拉很荣幸地将其慵懒奢华、贴心服务、探索体验的独特品牌带到亚洲和中东地区的更多绝美目的地，包括越南、巴厘岛、普吉岛和阿布扎比。

Development

The Anantara Experience was born in 2001, with the launch of the first Anantara Resort in Thailand's historic seaside resort enclave of Hua Hin. It sought to bring the guests closer to the heart of the rich culture and history of Thailand by surrounding them in the atmosphere of a traditional Thai village. It highlighted the experience with interactive immersions in the culture of the destination, through cooking courses, fruit carving demonstrations, a weekly floating market and even instruction in the traditional martial art of Muay Thai.

Since then, the Anantara Experience has been brought to Thailand's far northern Golden Triangle region, with the unique and world-renowned Anantara Elephant Camp experience, and to the island of Samui, offering an immersion in the rich culture of southern Thailand. In 2006, the Anantara Experience went global, with the launch of Anantara Maldives, providing guests with the opportunity to live the fantasy of a truly remote desert island getaway.

Inspired by this success and the positive feedback we have received from our guests worldwide, Anantara is proud to be bringing our unique brand of laid-back luxury, intuitive service and sense of discovery to more breathtaking destinations across Asia and the Middle East, including Vietnam, Bali, Phuket and Abu Dhabi.

酒 店 概 况

东方红树林安纳塔拉水疗酒店坐落于水岸，是喧嚣都市生活之巅的一处天然静居。作为综合式酒店、码头、店铺和住宅区域的一部分，其独特位置为宾客提供一次精彩的阿布扎比体验。该区域呈现异域美态的宝贵红树林作为完美背景，映衬着酒店壮观建筑所体现的丰富文化传统。在东方红树林安纳塔拉水疗酒店，探索阿联酋都市生活的豪华住宿定义，同时感受四处洋溢的天然魅力。

Overview

Majestically set on the waterfront, Eastern Mangroves Hotel & Spa by Anantara is a natural haven on the cusp of urban life. Forming part of an integrated hotel, marina, retail and residential destination, its unique location offers guests an exceptional Abu Dhabi experience. The exotic beauty of the region's precious mangroves provides the perfect backdrop for the rich cultural traditions embodied in the hotel's impressive architecture. Discover luxury accommodation definitive of urban Emirati life, yet filled with natural charm at Eastern Mangroves by Anantara.

酒 店 特 色

漫步宏伟的大厅和廊道、以及宽阔的中央通道，体验阿拉伯设计令人叹为观止的美感。一处诱人的世外桃源，与喧闹的市中心仿佛分隔成两个世界。

Feature

Wander through the majestic halls and corridors, a breathtaking showcase of Arabian design, with grand central passageways and arched ceilings. An inviting sanctuary blissfully sheltered from the hustle and bustle of the city centre.

酒店室内

在阿布扎比这家风景如画的酒店，超过200间、皆装饰以舒适别致的风格、洋溢着奢华的文化气息的客房可随心选择。在 Paychaylen 享受最新鲜美味的料理，或在 Impressions 轻松小酌，远方红树林美景更是锦上添花。在无与伦比的安纳塔拉水疗馆唤醒顾客的感官。在顶级健身房里恢复充沛体力，然后在凉亭遮阴的泳池畔放松休憩。无论商务或休闲，东方红树林都是一处充满宁静与优雅的完美所在。

Interior

Choose from over 200 rooms at this picturesque hotel in Abu Dhabi, all decorated in a comfortable yet chic style and boasting lavish cultural touches. Enjoy the freshest and finest cuisine at Paychaylen or relax for a drink at Impressions, complemented by views across the mangroves. Rejuvenate your senses at the incomparable Anantara Spa. Energise at the state-of-the-art gym and then relax poolside under the shade of a gazebo. For business or leisure, Eastern Mangroves provides a backdrop ripe with serenity and elegance.

酒 店 配 套

■ 宴会厅和六个分会场

可容纳多达 350 位客人，为宾客计划一次难忘的活动。配备天花板挂钩和多个视听连接，宴会厅拥有最为先进的设备，可进行复杂的演示和展览。为宾客提供进行一次难忘活动所需的一切。宴会厅旁设有温馨的休息区，可以在此休憩。

Services and Amenities

■ Ballroom and Six Breakout Rooms

Facilities for up to 350 guests enable you to plan memorable events and unforgettable occasions. With ceiling skyhooks and multiple audio visual connections, the ballroom features state-of-the-art equipment allowing for sophisticated presentations and displays for an impressive and unforgettable event. It adjoins the ballroom, kick-back in the cozy lounge area.

■ 休闲活动

东方红树林安纳塔拉水疗酒店提供各种休闲娱乐活动。在阿布扎比红树林美景映衬下的一处水岸绿洲，酒店充分利用这个背景，让宾客获得两全其美的探索体验。酒店专属码头衔接该区域最受欢迎的运河，非常适合航行、划艇、观赏鸟和野生动物等活动。此独特位置让东方红树林成为阿布扎比最好的酒店之一，非常适合举家户外游览和新鲜刺激的休闲活动。

■ 水疗中心

于阿布扎比充满活力的风景之间找寻一处舒适静逸的绿洲。以现代风格诠释传统的土耳其浴室，安纳塔拉水疗提供招牌安纳塔拉疗程、蒸汽和桑拿设施以及单独的男宾和女宾区域。在配有各种室内和户外设施的豪华理疗室和休息大厅中，体验水疗治愈力量为顾客带来的宁静。在宝石色的水疗间内享受点缀着现代豪华的古老手法。一处迷人宁静的环境，在阿布扎比豪华度假村之中特立独行。

■ 健身

宽敞的健身房为客人提供各种锻炼课程以及心血管和肌肉训练设备。设有单独的女性健身区域，顶级设施配备最新技术，包括多个视听连接。锻炼课程之间需要休息时，宾客可在健身房内的果汁吧享受一杯健康饮料。女士专用健身房位于同一楼层。

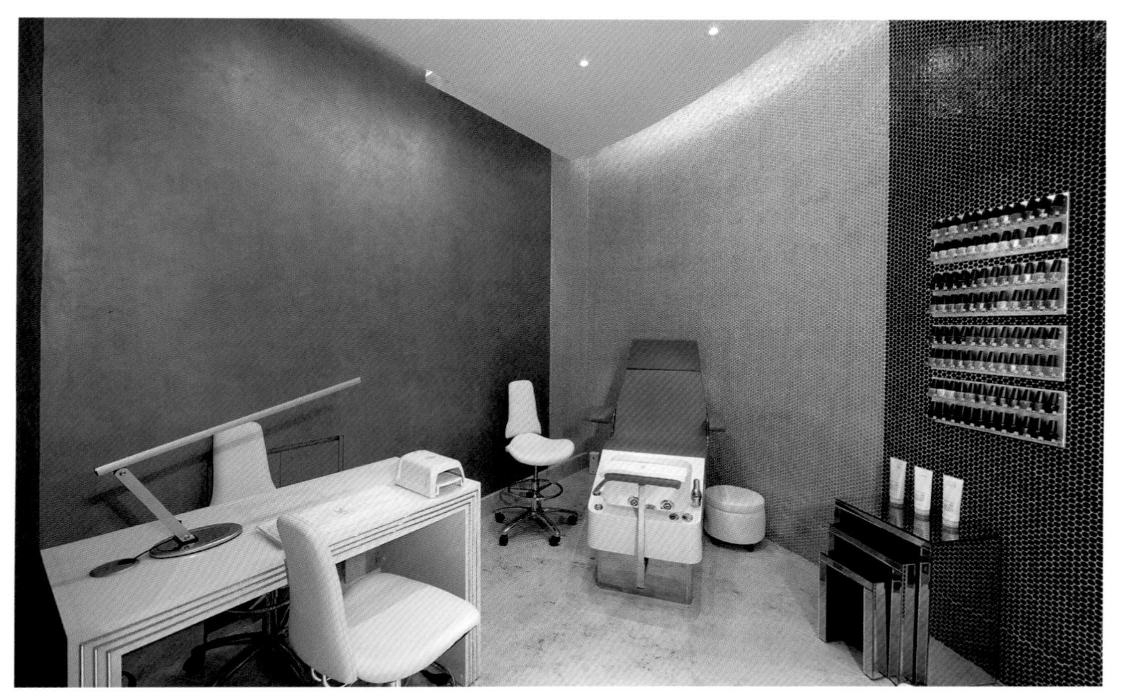

■ Entertainment

Eastern Mangroves Hotel & Spa by Anantara provides an assortment of activities to relax and play. A waterfront oasis in the city, with views over the region's mangroves, the Abu Dhabi luxury hotel makes the most of its setting to enable guests to explore the best of both worlds. An on-site marina provides access to the area's popular canals, which are ideal for sailing, kayaking, bird watching and wildlife activities. This unique location makes Eastern Mangroves one of the best Abu Dhabi hotels for family friendly outdoor excursions and exciting leisure pursuits.

■ Spa

Amid Abu Dhabi's vibrant landscape, find an oasis of comfort and calm. A modern twist on the traditional Turkish Hammam, the Anantara Spa features signature Anantara treatments, steam and sauna facilities, as well as separate male and female areas. Experience the tranquillity of therapeutic wisdom in decadent treatment rooms and relaxation lounges with a choice of indoor or outdoor facilities. Indulge in ancient practices mixed with modern luxury amid the jewel toned spa; an environment both mesmerizing and serene, unique among Abu Dhabi luxury resorts.

■ Fitness Center

The expansive workout room provides guests with a diverse range of exercise classes, along with cardiovascular and muscle training equipment. With separate workout spaces for women, the state of the art facilities feature the latest technology including multiple audio visual connections. An on-site juice bar offers healthy treats to be enjoyed when feel like taking a breather from exercise session. A dedicated ladies-only gym is located on the same level.

Jumeirah at Etihad Towers | 卓美亚阿提哈德中心酒店

Keywords 关键词
Dramatic Sculpture 巨型雕塑
Complete Facilities 丰富配套
Luxury Experience 奢华体验
Arabian Touch 阿拉伯风情

品牌链接

卓美亚酒店堪称世界上最奢华、最具创新意识的酒店，已荣获无数国际旅游奖项。集团始建于 1997 年，志存高远，立志通过打造世界一流的奢华酒店及度假酒店，成为行业领袖。在这些辉煌成就的基础上，卓美亚集团 (Jumeirah Group) 于 2004 年成为迪拜控股旗下一员，在这艘汇聚了迪拜众多项尖企业和项目的航母上，卓美亚驶向了成长与发展的新旅程。

About Jumeirah Hotels and Resorts

Jumeirah Hotels & Resorts are regarded as among the most luxurious and innovative in the world and have won numerous international travel and tourism awards. The company was founded in 1997 with the aim to become a hospitality industry leader through establishing a world class portfolio of luxury hotels and resorts. Building on this success, in 2004 Jumeirah Group became a member of Dubai Holding – a collection of leading Dubai-based businesses and projects – in line with a new phase of growth and development for the Group.

酒店地址：阿联酋阿布扎比市滨海西路 225 号
传　　真：+971 2 6444214

Address: PO Box 225, West Corniche, Abu Dhabi, United Arab Emirates
Fax: +971 2 6444214

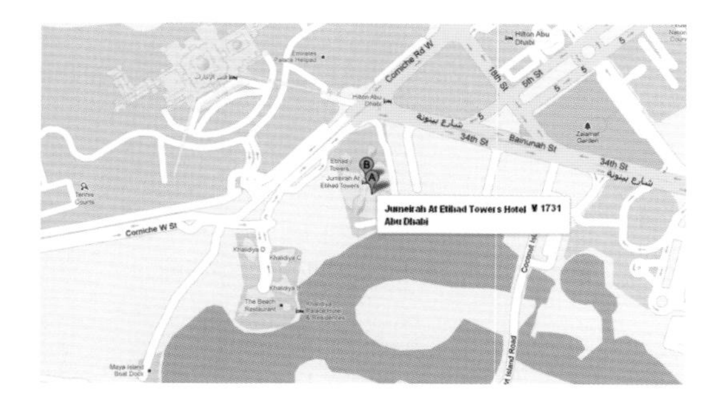

酒店开发管理

卓美亚 Etihad Towers 是 Sheikh Suroor bin Mohammed Al Nahyan 殿下资助，与建筑公司 DBI Australia 于 2006 年携手打造兴建。

Development and Management

Jumeirah at Etihad Towers was developed under the patronage of His Highness Sheikh Suroor bin Mohammed Al Nahyan's Projects Department with architects DBI Australia starting work on this unique project in January 2006.

酒店概况

卓美亚 Etihad Towers 位于阿拉伯联合酋长国的首都阿布扎比西边沿海的 Ras Al Akhdar 地区。距阿布达比国际机场 37 km。

Overview

Jumeirah at Etihad Towers is located on the shores of the Arabian Gulf in the exclusive Ras Al Akhdar area, on the West Corniche of Abu Dhabi, 37 km from Abu Dhabi International Airport.

酒店外观

该海滨建筑是由五座大楼组成，当中包括 1 幢 280 m 高的豪华酒店（卓美亚 Etihad Towers 酒店）、1 幢出租面积达 46 000 m^2 的办公室大楼及观景台，及 3 幢共 885 个单位的公寓大楼。5 幢大厦的四楼平台互相连接，设有时尚的餐厅及超过 30 间高级时装精品店。

大楼外形如巨型雕塑，与现代精致的室内设计和富有文化色彩的装饰完美配合。Sheikh Suroor bin Mohammed Al Nahyan 殿下特别捐出私人珍藏品，为酒店的公共空间及会议预备区域增添一份本土的阿拉伯风情。

Exterior

The iconic beachfront project will feature five towers, including a 280 m tall luxury hotel (Jumeirah at Etihad Towers), a prestigious office tower with 46,000 sqm of leasable office space and an observation deck, and three towers housing 885 apartments. The connecting four-level Podium will feature contemporary restaurants and over 30 premium boutiques.

The dramatic sculptural forms of the towers compliment the sophisticated and contemporary interior design, with a cultural edge. Art works from His Highness Sheikh Suroor bin Mohammed Al Nahyan's private collection provide a local Arabian touch in public areas and conference pre-function areas.

酒 店 室 内

卓美亚 Etihad Towers 由 382 间客房及套房、199 间服务式住宅及 12 间餐厅、酒廊与酒吧组成。酒店同时提供一个私人沙滩、泰丽丝水疗中心、泳池、健康中心及一个共 13 间会议室的会议中心和 Mezzoon 大型宴会厅。

Interior

Jumeirah at Etihad Towers is made up of 382 guest rooms and suites, 199 serviced residences and 12 restaurants, bars and lounges. The hotel also offers a private beach, Talise Spa, pool, health club and extensive conference centre with 13 meeting rooms and the grand Mezzoon Ballroom.

酒店配套

■ 公寓

199 间全服务式公寓位于第 2~25 层，面积最小的在 57 m^2，另有两至三房共 170 m^2 的宽敞单位。所有公寓均拥有令人无可抗拒的壮阔海景和独立的接待处。

■ 餐厅酒吧

卓美亚 Etihad Towers 设有 12 间餐厅及酒廊，提供各式各样的美食选择。位于平台的海景餐厅包括：别致的黎巴嫩餐厅 Li Beirut 和提供全日豪华餐饮服务的 Rosewater，两间餐厅都附设露台，提供户外用餐选择。另外，餐饮选择还包括传统法国餐厅 Brasserie Angélique，时尚日式居酒屋 Tori No Su，轻松的海边餐厅 Nahaam 及位于沙滩尽头的海鲜餐厅 Scott's。

位于室外泳池边的 The pool bars 提供一系列的果汁、冰沙及其他饮品和点心小吃供客人选择。另外，酒店还设有位于 62 楼装潢豪华的 Ray's Bar 和位于 63 楼提供精致亚洲风味美食的餐厅 Quest。两间餐厅的顾客都可以尽情饱览阿布扎比壮丽的海滨及市内景致。水疗中心咖啡厅则提供和谐宁静的气氛及精美茶点，而旅客亦可选择于酒店的大堂酒吧廊尽情放松。

Services and Amenities

■ Residences

The 199 fully-serviced residences located on Levels 2 – 25, vary from studios of 57 sqm to spacious one, two and three bedroom apartments of up to 170 sqm. All inspirationally designed residences offer irresistible sea views and share an exclusive reception area.

■ Restaurants & Bars

Jumeirah at Etihad Towers offers a choice of 12 restaurants, bars and lounges with a variety of diverse culinary options. The collection of sea view restaurants on The Podium include Li Beirut, a chic Lebanese restaurant and Rosewater, a luxurious all-day dining restaurant, both with outdoor terrace seating; Brasserie Angélique, a classic French brasserie; Tori No Su, a stylish Japanese restaurant and lounge; Nahaam, a casual beach restaurant; and Scott's Seafood Restaurant, located on the water off the beach.

The pool bars on the outdoor pool terrace offer a selection of juices, smoothies and other drinks as well as a menu of light bites and snacks. Ray's Bar is a lavishly appointed bar located on Level 62. Quest, an exquisite pan-Asian signature restaurant, is housed on Level 63. Both levels provide guests with stunning views of the Corniche and the Abu Dhabi cityscape, offering guests a sophisticated dining experience with views to match. The spa café Bhodi offers light and refreshing cuisine in a soothing and calm atmosphere. Guests can also relax and indulge in the hotel's Lobby Lounge and Lobby Bar.

RENOIR

■ 客房

卓美亚 Etihad Towers 酒店共有 66 层，其 382 间具有当代豪华设计风格和配置的客房及套房，分布于大楼的 27~60 层之间。所有的房间面积介乎 40~60 m^2，均能饱览令人惊叹的海景及享受完善生活科技配套设施。所有行政大楼客房、豪华行政客房、高级豪华行政客房、行政套房、阿提哈德套房、空中套房及阿提哈德皇家套房的住客都可享有特别的行政服务及免费使用拥有 360 度落地玻璃窗景观的行政楼层休息室。

■ 会议及宴会设施

酒店四楼设有阿布扎比最大的会议及宴会中心，附设 13 个会议室及豪华的 Mezzoon 宴会厅。该宴会厅可举办容纳 1 400 位宾客的会议，或 1 000 位宾客的晚宴，亦可分成 4 个独立的会议室。中心亦设有一个宽阔的预备区域，这里有绿色客房 (green room) 及商务中心，再加上特别设计、足以运载私家车的大型升降机。

■ Rooms

Jumeirah at Etihad Towers offers guests 66 floors with 382 beautifully appointed guest rooms and suites which are located between Levels 27 to 60, each with luxurious contemporary décor and furnishings. All rooms offer stunning sea views and range from 40 sqm to 60 sqm in floor area with the latest fully integrated lifestyle technology. Guests staying in the Tower Club, Deluxe Club, Grand Club Rooms and Club, Etihad, Sky and Royal Etihad Suites receive exclusive service and access to the Club Executive Lounge. This gorgeously appointed lounge offers guests 360 degree views with floor to ceiling windows.

■ Conference & Banqueting Facilities

Extensive conference & banqueting facilities are provided through the hotel's purpose built Conference Centre, located on Podium Level 4. The Conference Centre is among the largest in Abu Dhabi and features 13 meeting rooms and the stunning Mezzoon ballroom. Mezzoon can comfortably accommodate up to 1,400 guests for a conference or 1,000 guests for dinner, and can be divided into four separate sections. The centre also offers spacious pre-function areas as well as a green room and business centre facilities. One of the unique elements of the ballroom is a special service elevator with the capacity to lift cars.

■ 泰丽丝水疗中心

卓美亚 Etihad Towers 的水疗中心设于平台 3M 楼层，附有 13 个私人疗程套房，包括 Rasou 套房（以传统中东方法，结合泥浆、草本蒸气及精油混合的暖水淋浴方式进行深层清洁及排毒）和 Hammam 套房（以传统的热云石床、私人蒸气桑拿房配合温差和按摩，帮助身体去除角质，恢复活力）。

■ 运动及休闲

卓美亚 Etihad Towers 的 6P 健身中心位于平台 3M 楼层，邻近私人沙滩，设有高科技的健身仪器。酒店亦设有花园及水池，为喜欢静态活动的旅客提供宁静的空间。另外，相连的平台也进驻了超过 30 间时装精品店，为旅客提供豪华购物体验。

■ Talise Spa

The Talise Spa at Jumeirah at Etihad Towers is located on Podium Level 3M and is comprised of 13 private treatment suites including a Rasoul suite (traditional Middle Eastern method of cleansing and detoxifying, using mud, herb-scented steam and an oil-infused warm shower) and the Hammam suite (a traditional heated marble goebektas bed, private steam and sauna room combining exposure to different temperatures, invigorating body exfoliation, steam immersion and rejuvenating massage).

■ Sports & Leisure

Jumeirah at Etihad Towers' state of the art 6P Gym is equipped with TechnoGym equipment on Podium Level 3M, next to Talise Spa, as well as a private beach. For guests looking for lower impact recreation, a landscaped oasis of pools and gardens provide a tranquil space. Etihad Towers also houses a selection of high end shopping with over 30 premium boutiques located in the Podium.

Jumeirah Bilgah Beach Hotel | 卓美亚彼耳加海滩酒店

Keywords 关键词
Modern Style 现代风格
Retreat and Vacation 归隐度假
Artistic Atmosphere 艺术气息
Tranquil and Relexing 宁静放松

酒店地址：阿塞拜疆共和国巴库市彼耳加区 AZ1122 Gelebe 大街 94 号
电　　话：+ 994 12 565 4000

Address: 94 Gelebe Street, Bilgah District AZ 1122, Baku, Azerbaijan
Tel: + 994 12 565 4000

品牌链接

卓美亚酒店堪称世界上最奢华、最具创新意识的酒店，已荣获无数国际旅游奖项。集团始建于 1997 年，志存高远，立志通过打造世界一流的奢华酒店及度假酒店，成为行业领袖。在这些辉煌成就的基础上，卓美亚集团 (Jumeirah Group) 于 2004 年成为迪拜控股旗下一员，在这艘汇聚了迪拜众多顶尖企业和项目的航母上，卓美亚驶向了成长与发展的新旅程。

About Jumeirah

Jumeirah Hotels & Resorts are regarded as among the most luxurious and innovative in the world and have won numerous international travel and tourism awards. The company was founded in 1997 with the aim to become a hospitality industry leader through establishing a world class portfolio of luxury hotels and resorts. Building on this success, in 2004 Jumeirah Group became a member of Dubai Holding – a collection of leading Dubai-based businesses and projects – in line with a new phase of growth and development for the Group.

酒店概况

卓美亚彼耳加海滩酒店位于里海沿岸300 m宽的彼耳加海滩上，这是阿塞拜疆的第一座国际性豪华城市度假酒店。酒店的位置在距离繁市中心较近的范围内，是一座集休闲与商务功能于一身的终极田园式酒店。酒店内的176间客房均可以欣赏到里海沿岸美丽的海景风貌。

这座完全现代风格的酒店由176间豪华套房和14座独栋别墅组成，设有多种设施可满足商务和休闲需要。酒店内还设有多个别致的餐厅和酒吧，会议中心，多功能厅和私人会议室，水上乐园，私人海滩，保健水疗会所，网球场、足球、排球和篮球场地，保龄球馆以及夜总会等。卓美亚彼耳加海滩酒店为游客提供一个Absheron半岛归隐度假的好去处，近郊的地理位置也成为酒店的亮点之一。

Overview

Set alongside a 300 meter Bilgah Beach on the Caspian Sea shore, Jumeirah Bilgah Beach Hotel is the first international luxury city resort in Azerbaijan. This resort is the ultimate idyllic retreat within close range of the city's major attractions. Each of the 176 guest rooms and suites offer sweeping sea views over the Caspian Sea.

This contemporary lifestyle complex comprises 176 luxurious rooms and suites, and 14 stand–alone cottages along with extensive business and leisure facilities.The hotel features an array of exceptional restaurants and bars, conference centre, multi–function venue and private meeting rooms, water park, private beach, a health club & spa, tennis court, football, volleyball and basketball fields, bowling alley, and a night club. Jumeirah Bilgah Beach Hotel offers the ultimate retreat on the Absheron Peninsula within close proximity to the major attractions of Baku and its suburbs.

酒 店 特 色

卓美亚彼耳加海滩酒店是巴库的第一个国际性旅游度假胜地，这里是体验正宗的阿塞拜疆热情的理想去处，在此游客可以感受到卓美亚酒店品牌的标志性奢华服务。

Feature

Jumeirah Bilgah Beach Hotel is Baku's first international luxury resort, serving as the ideal retreat for those seeking to experience authentic Azeri hospitality while enjoying Jumeirah's signature standards of luxury.

酒店室内

酒店大厅迎面摆设着一系列独特的设计元素：宫殿般的灯光设计，成千上万、大大小小、形状各异的金属球经过精心设计编排使墙面变得极其富有艺术感，现代感浓郁的金属碎片和手工玻璃碎片。在前厅的正中悬挂着 58 m 长的巨大枝形吊灯，上面装有 72 000 个 LED 灯泡。此外大厅内还设有温馨的休息区、台球室、购物街、钢琴酒吧、以及一座陈列着艺术、历史、摄影、建筑、设计等科目中珍贵书籍的图书馆。

Interior

The Hotel's lobby areas represent an array of unique design elements: palatial lighting features, walls decorated with meticulously detailed artful configuration of thousands of metal balls of different shapes, variety of modern design pieces from metal and hand-blown glass. At the center of the atrium lobby is gold and silver plated 58-meter long grand chandelier with 72,000 LED lights. The lobby area features a number of cozy seating areas, a billiards room, shopping arcade, a piano lounge and a unique library with rare collectible books of art, history, photography, architecture, design and other.

Jumeirah
STAY DIFFERENT™

酒店配套

■ 会议中心

酒店内设有大型的会议中心，总面积达 1 500 m^2，包含彼耳加宴会厅，Mardakan 套房和叠层，巨大的多功能厅以及三个紧靠综合商务中心的私人会议室。所有的会议区都配有最先进的科技设备，提供免费的 Wi-Fi 连接。酒店非常适宜举办商务会议和特殊的社会活动。

■ 美食体验

卓美亚彼耳加海滩酒店的与众不同之处在于它为客人提供多种多样的美食体验，在酒店各处客人都可以享受到别具特色的餐饮和景观。酒店内共设有 10 座餐厅，分别适合家庭、非正式会议用餐以及正式用餐等。在这里客人总能找到一处适宜与家人朋友吃简单午餐、喝下午茶或品尝海鲜的的角落，同时还可以欣赏到里海沿岸的独特风光。

Services and Amenities

■ Dining

Jumeirah Bilgah Beach Hotel differentiates itself by the large and varied culinary experiences available throughout the complex. Different flavors in ten restaurants are available catering for family, informal meeting and formal dinner. People could always find places to taste light lunch, afternoon coffee, sea food etc. with friends while enjoying the stunning scenery of the Caspian Sea.

■ Conference Center

Jumeirah Bilgah Beach Hotel Baku features an extensive Conference Centre with a total area of 1,500 m^2 including the grand Bilgah Ballroom; Mardakan Suite & Terrace, a large multipurpose venue; and three private boardrooms located next to a comprehensive Business Centre. All spaces incorporate the most advanced technology, provide complimentary Wi-Fi and are easily adaptable for a business meeting or conference or a spectacular social event.

■ 泰丽丝水疗中心

在酒店的正中奢华掩映下的泰丽丝水疗中心使客人感到宁静放松。所有的客人都会得到专属治疗专家的指导。水疗区域占地 500 m^2，提供面部、身体等的按摩治疗服务。在这里，客人可以体验到泰式、巴厘岛式以及传统的土耳其蒸气浴疗法。

■ Talise Spa

Embedded in the luxurious surroundings of the Jumeirah Bilgah Beach Hotel is the tranquil atmosphere of Talise Spa. All guests are welcomed by therapists attentive to personal needs. The Spa occupies 500 m^2 of space and offers a full range of facial, body and massage treatments featuring a unique blend of Thai and Balinese treatments and the traditional Turkish bath experience.

■ 休闲娱乐

卓美亚彼耳加海滩酒店内的水上乐园是区域内最大的，游客可以在这里享受到各种各样的水上乐趣。勇敢者可以挑战惊险刺激的急速划艇，而喜欢休闲放松的游客则可享受室外泳池提供的喷泉和阳光甲板带来的乐趣。

■ Leisure and Entertainment

Jumeirah Bilgah Beach Hotel's waterpark is the largest in the area and offers attractions and fun for everyone. The daredevils will want to try the breathtaking speed slides, while those looking for refreshing relaxation will enjoy the numerous interconnected outdoor swimming pools featuring fountains and sun decks.

Keraton at The Plaza, Jakarta | 雅加达苏丹皇宫广场酒店

Keywords 关键词

Refined 精致

Business & Leisure 商务休闲

Iconic 地标

Indonesian Style 印尼风格

品牌链接

雅加达苏丹皇宫广场酒店，隶属喜达屋酒店与度假村集团旗下的豪华精选系列。作为喜达屋酒店与度假村国际集团旗下酒店，豪华精选（Luxury Collection®）是一系列可提供独特地道体验，为旅客留下珍贵难忘回忆的精选酒店。豪华精选致力于为各地旅游爱好者们打开一扇通往全球最激动人心和最令人向往目的地的大门。 每家酒店及度假村均独具特色、风情各异；堪称旅行目的地门户的豪华精选酒店将使旅客尽情领略原汁原味的当地文化的无限魅力。精美的装饰、豪华的布置、无与伦比的服务和先进的现代化便利设施将为旅客打造独特而丰富的入住体验。

About Keraton

Keraton is one of the Luxury Collection® hotel brands owned by Starwood Hotels & Resorts. As a subsidiary of Starwood Hotels, The Luxury Collection® is a selection of hotels and resorts offering unique, authentic experiences that evokes lasting, treasured memories. For the global explorer, The Luxury Collection offers a gateway to the world's most exciting and desirable destinations. Each hotel and resort is a distinct and cherished expression of its location; a portal to the destination's indigenous charms and treasures. Magnificent décor, spectacular settings, impeccable service and the latest modern conveniences combine to provide a uniquely enriching experience.

酒店地址：印度尼西亚雅加达 Jl. MH. Thamrin Kav. 15
电　　话：(62)(21) 5068 0000

Address: Jl. MH. Thamrin Kav. 15, Jakarta, Indonesia.
Tel: (62)(21) 5068 0000

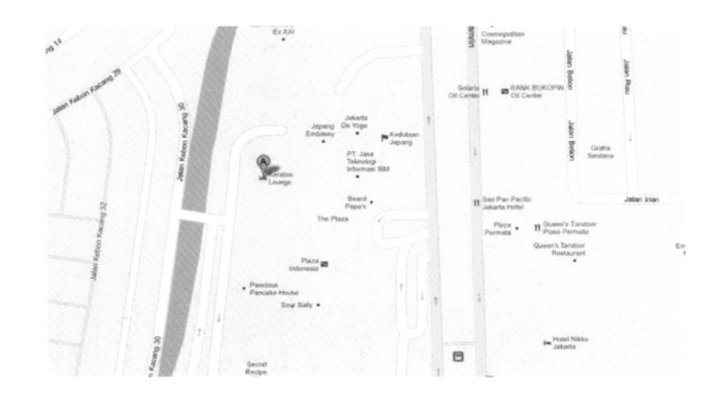

MUSEUM

酒店概况

雅加达苏丹皇宫广场酒店位于雅加达繁华的中央商务区，是这座活力都市中的一处精致幽居之所。酒店与一座高档购物商场紧密相连，并毗邻企业办事处和旅游景点，如城市的地标性建筑——欢迎雕像。

“Keraton”在爪哇语中是“宫殿”的意思。酒店设计灵感来自于昔日的皇家宫殿遗迹，以现代风格加之无与伦比的设计来重新诠释这座豪华酒店。

Overview

Located in a worldly business district, Keraton at The Plaza is a refined destination in this vibrant city. The hotel is connected to an opulent shopping mall and close to offices and attractions, such as the iconic Selamat Datang statue.

Meaning “Palace” in Javanese and inspired by commissioning relics from the royal palaces of yesteryear and reinterpreted with contemporary design, exception surpasses expectation at Keraton at The Plaza, a Luxury Collection Hotel, Jakarta.

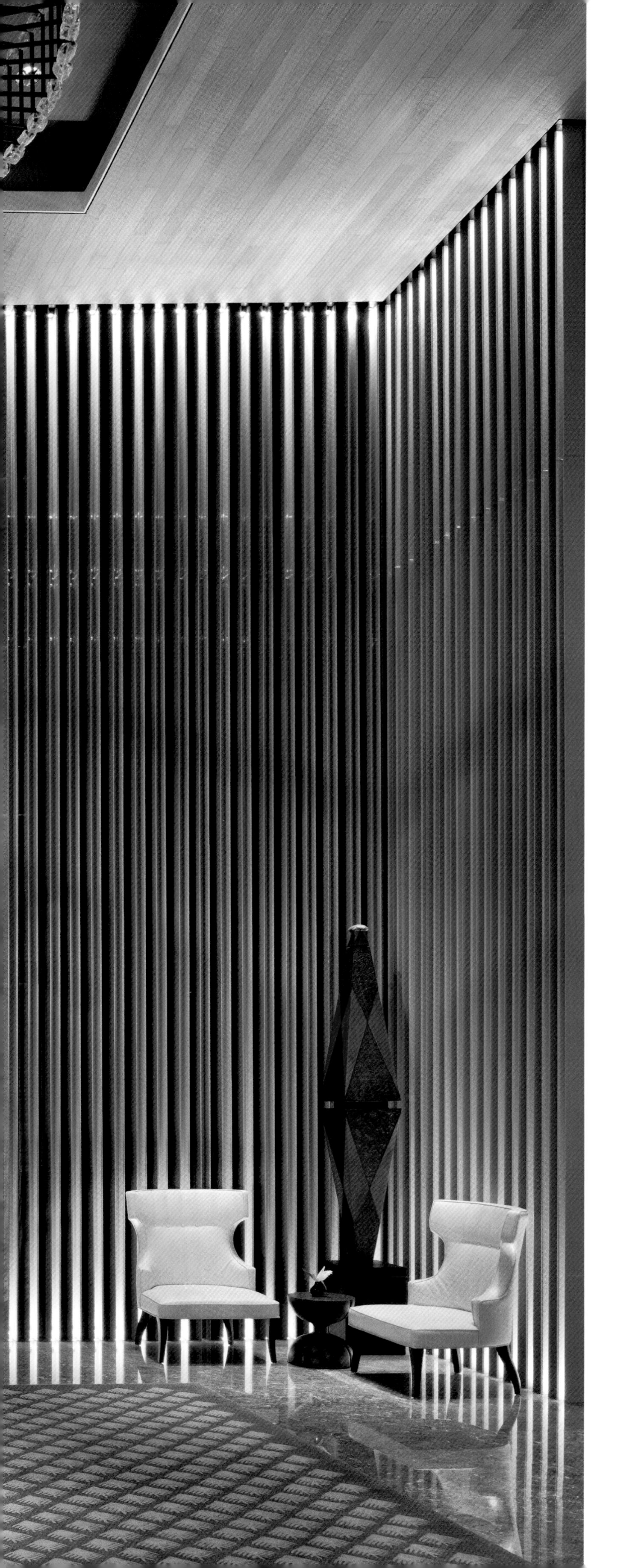

酒 店 特 色

在这里，宾客将能领略到独特鲜明的热情、无可挑剔的服务以及最先进的现代化便利设施，感受到印尼久负盛誉的优雅传统和盛情款待。

Feature

Discover a definitive hospitality experience with impeccable service and the latest in modern conveniences and amenities.

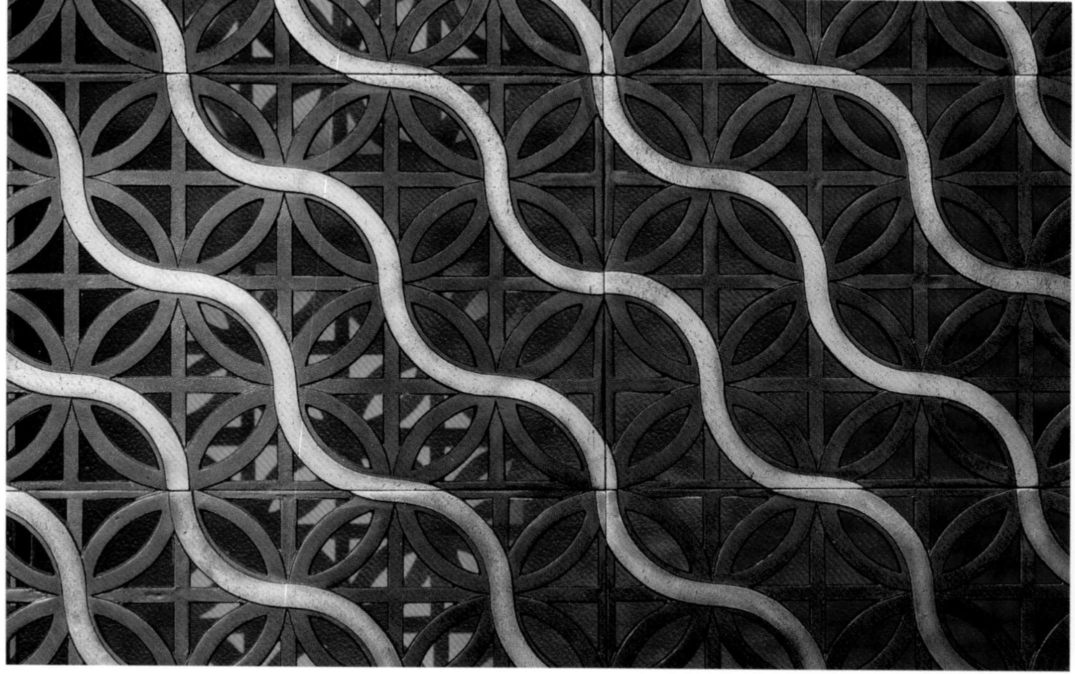

酒 店 室 内

酒店所有的 140 间客房与套房均采用现代化的印尼装饰风格。

Interior

All 140 guest rooms and suites have been sumptuously appointed in contemporary Indonesian style.

酒店配套

■ 餐饮

酒店非凡卓越的餐厅提供具有浓郁地方风味的美食。在豪华的餐饮环境中，每一种菜肴都将带给宾客美妙独特的味觉体验。

■ 会议

雅加达苏丹皇宫广场酒店位于中央商务区的黄金地段，紧邻印尼广场，是举办高端商务活动或豪华社交聚会的非凡场所。装潢华丽的行政会议室可轻松容纳6人，且配有先进一流的高科技设施，自然采光十分充足。

Services and Amenities

■ Dining

The exceptional dining venues entwine gourmet cuisine with indigenous flavor. Combined with sumptuous surroundings, a unique, culinary experience will be enjoyed with every dish.

■ Meetings

Located in the prime locations of the central business district, next to Plaza Indonesia, Keraton Hotel is extraordinary place to hold high-end business or luxury social gatherings.

The magnificently decorated executive conference room can comfortably accommodate 6 people and equipped with state-of-the-art high-tech facilities with adequate natural lighting.

■ 客房及套房

酒店共140间客房及套房，配有精美的床上用品和华丽的浴室。一系列贴心周到的客房设施及非凡卓越的全天候服务将为顾客带来完美无瑕的住宿体验。

■ Rooms & Suits

All 140 guest rooms and suites have been sumptuously appointed in contemporary Indonesian style, with exquisite bedding and palatial bathrooms. An array of thoughtful features and exceptional, around-the-clock service provide a seamless experience.

Fairmont Jaipur, India

印度斋浦尔费尔蒙酒店

Keywords 关键词

Royal Style 皇家气派

Exotic Flavor 异域风情

Traditional Elegance 传统高雅

品牌链接

费尔蒙凭借其独具特色的酒店系列和享誉全球的卓越声誉在全球酒店业始终占据领先地位。其多元化的酒店组合中包含了众多著名地标建筑、高档度假村以及位于现代都市中心的卓越酒店。从独具风情的夏威夷海滩和百慕大群岛到热闹繁华的纽约市中心，所有酒店均可为前来下榻的尊贵来宾提供独一无二的“费尔蒙”客户体验。

About Fairmont

Fairmont is a leader in the global hospitality industry, with a distinctive collection and a worldwide reputation for excellence. Its diverse portfolio includes historic icons, elegant resorts and modern city center properties. From the beaches of Hawaii and Bermuda to the heart of New York City, all its hotels offer a superior guest experience that is uniquely “Fairmont”.

酒店地址：印度斋浦尔 Riico Kukas 大道 2 号
电　　话：+91 142 642 00 00

Address: 2, Riico Kukas, Jaipur India
Tel: +91 142 642 00 00

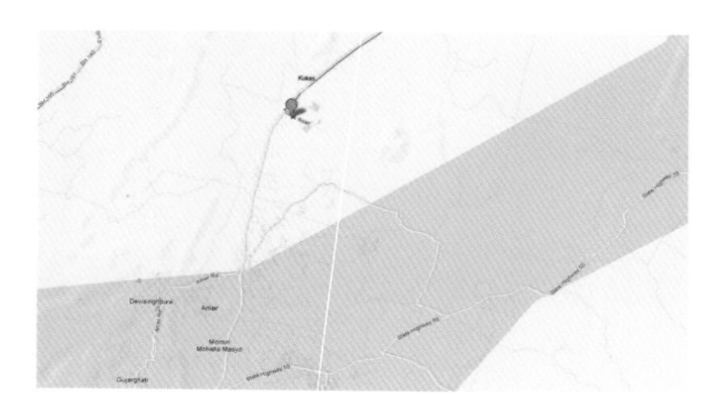

酒 店 概 况

斋浦尔，又被称为“粉红城市”，是印度拉贾斯坦邦首府，也是开放的世界旅游城市。斋浦尔费尔蒙酒店的设计灵感来自于蒙兀儿王朝昔日的皇家拉其普特，其令人振奋的建筑和装饰，是对粉红城市的颂歌。这家豪华酒店，坐落在雄伟的阿拉瓦利山脉之中，共有255间客房和套房，室内高雅而传统的拉贾斯坦装饰和现代化的设施完美结合。

Overview

Jaipur, the “Pink City” and the stunning capital of the Indian state of Rajasthan, is a hub for tourists and visitors from all over the world. Fairmont Jaipur is an ode to the pink city through its awe inspiring architecture and decor, inspired by the Mughal dynasty and Royal Rajputs of yore. This luxury Jaipur hotel, is nestled amongst the majestic Aravalli hills, all 255 rooms and suites are elegantly appointed, to reflect a perfect blend of traditional Rajasthani d é cor and modern amenities.

酒店特色

酒店设计以蒙兀儿帝国风格为主，并将蒙兀儿时代的皇宫重新呈现出来。

Feature

The Mughal–dynasty–style hotel represented the ancient Mughal dynasty palace .

酒店配套

■ 餐饮

ZOYA 餐厅充满异域风情，特别强调正宗的拉贾斯坦美食。迷人的 ANJUM 餐厅，为费尔蒙的国际茶饮酒文化庆祝活动加上了印度的本地风味。优雅的 AZA 鸡尾酒酒吧则是个安静的阅读场所。

■ 会议

斋浦尔费尔蒙五星级酒店作为商业事务会议之选是独一无二，无以伦比的。酒店内设有一个专门的会议中心，推出理想的顶级服务产品。可举办晚宴或盛大婚宴的宽敞草坪，配合全市最大的会议厅和 5 个风格独特的全国最先进的会议室，以提供最好的休闲和商务服务为目的，让人沉浸在宁静和优雅的气氛之中。

Services and Amenities

■ Dining

The opulent accommodation is complemented with an array of exotic dining options, from the vibrant and energetic ZOYA, all day dining restaurant, with a special emphasis on authentic Rajasthani cuisine, the charming ANJUM, where Fairmont's internationally celebrated tea drinking culture comes to India with an exciting local twist, to the elegant AZA, our library bar for a quiet evening of aperitifs and signature cocktails.

■ Meetings

Conduct business affairs at a location that is unique and unparalleled in beauty and charm. This Jaipur 5 star hotel has a dedicated convention centre, ideal for premier product launches, gala dinners or grand weddings. With spacious lawns complementing the city's largest convention hall and ably supported with five individually styled, state-of-the-art meeting rooms, Fairmont Jaipur encompasses the best of leisure and business, immersed in an atmosphere of tranquility and elegance.

■ 温泉水疗

在华丽的 Nirvana 温泉享受宁静。从按摩和水疗护理、冥想和瑜伽，Nirvana 提供全方位的服务，以帮助顾客放松心情，消除出差或旅行的疲劳。客人可以享受传统的阿育吠陀、异国情调的芳香和正宗的东方疗法。

■ 活动

年轻的游客在企鹅俱乐部和儿童游泳池可以享受热闹的活动。企鹅俱乐部提供一系列的室内和室外游戏和教育训练。斋浦尔酒店有一间游戏室，一个 42 个座位的电影院和一个室外游泳池很适合家庭旅客。

酒店的体育活动包括迷你高尔夫球场、门球、羽毛球、网球、大象马球和板球。室内游戏包括台球、棋牌、乒乓球等各种棋类游戏。

■ Spa

Tranquility and peace are provided at the beautiful Nirvana Spa. From massages and spa treatments to meditation and yoga, Nirvana offers a full range of services to help relax and forget the fatigue of business or travel. Guests can discover traditional Ayurveda, exotic aromatherapy, and authentic Eastern therapies. The resident bathologist can create a customized bath in the private environment of villas.

■ Activities

Young visitors might enjoy the lively activities at Penguin Club and the children's swimming pool. The Penguin Club offers an array of indoor and outdoor games and educational exercises. For families, Le M é ridien Jaipur features a game room, a 42-seat cinema, and an outdoor swimming pool.

Sporting activities include miniature golf, croquet, badminton, tennis, elephant polo, and cricket. Indoor games include snooker, chess, table tennis, and variety of board games.

The Okura Prestige Taipei

台北大仓久和大饭店

Keywords 关键词

Best Luxury 顶级奢华

Modern Classic 现代经典

Japanese Style 日式风格

酒店地址：中国台湾台北市南京东路一段9号
电　　话：+886(2)2523-1111

Address: 9 Nanjing E. Rd., Sec. 1, Taipei, Taiwan
Tel: +886(2)2523-1111

品牌链接

台北大仓为大仓品牌Prestige系列超五星级酒店。日本大仓饭店集团（Hotel Okura Co., Ltd）是日本当地最大的国际五星级连锁饭店集团，旗下的日本大仓饭店（Okura Hotels & Resorts）品牌，目前全球共有约25间大仓饭店。日本大仓饭店集团同时拥有日航饭店品牌（JAL Hotels），旗下包括Nikko Hotels International与Hotel JAL City二品牌。秉持“最好的ACS:极致的住宿、美食与服务经验”(The Best ACS :Best Accommodation, Best Cuisine, Best Service)为经营核心价值，日本大仓饭店集团旗下品牌在日本与海外拥有共约100间饭店。

大仓饭店以日本文化的守护者自居，自1958年创设以来，拥有50多年的悠久历史，与东京帝国饭店、东京新大谷饭店并称代表东京当地三大顶级饭店的“御三家”。在日本当地，大仓不仅被公认为最豪华的顶级饭店，更是最能表现与代表日本传统文化精神的国际五星级饭店，在日本人心目中占有非常重要的指标性的地位。大仓饭店的特色在于将日式的亲切细腻服务结合西方顶级饭店设备的便捷高效，承袭源自东京大仓以“亲切”与“和”的服务精神，延续大仓五十年细腻款待贵客之道。

About Hotel Okura

Hotel Okura Co., Ltd is the hotel chain managed by Hotel Okura which offers premium hotel services. The mission of this international group is to develop hotels that provide genuinely relaxing accommodation for guest worldwide by marrying Okura's deeply rooted sense of Japanese refinement with Westin functionality.

With Hotel Okura Tokyo serving as flagship hotel, Hotel Okura Co., Ltd operates 15 hotels in Japan and 7 around the world, and is working to expand mileage program with numerous airline companies, sales and reservation networks. Hotel Okura Co., Ltd entered into strategic joint marketing alliance with Banyan Tree Hotels & Resorts in October 2006, with Taj Hotels Resorts and Palace in September 2007 and formed a partnership with JAL Hotels in September 2010, toward launching combined reservation services, and membership privileges and points programs with the two groups. In this way, it has established a system to provide greater customer convenience.

Since established in 1958, Hotel Okura Tokyo, the flagship hotel of Hotel Okura Co., Ltd, has been recognized as one of the world's premier luxury hotels. By consistently providing the highest quality hospitality services, further enhanced by the unique characteristics, sensibility, traditions and culture of Japan, the hotel has earned the loyalty of elite travellers from across the world including previous presidents of the United States and former French President Jacques Chirac. In 1964, Hotel Okura Tokyo was selected as the first in Asia to host an international conference of the International Monetary Fund (IMF). The hotel quickly established a pre-eminent position with its outstanding service within just a few years of its founding.

The Okura

酒店特色

酒店以现代经典为风格走向，在简洁的颜色与线条的现代空间中，展现低调沉稳的氛围；每间客房皆备有日式机能卫浴设备；酒店餐厅提供精细无比又余韵无穷的正统日本料理。

Feature

The Okura Prestige Taipei is featured in contemporary design; each room is equipped with separate dry-wet bathing facilities; the Teppanyaki area allows guests to enjoy the flavors of fine meats and seafood cooked in another traditional way.

酒店室内

大仓久和大饭店客房由香港著名的饭店设计公司Chhada, Siembieda & Associates Ltd. 打造，以现代经典为风格走向，在简洁的颜色与线条的现代空间中，展现低调沉稳的氛围。自然柔和的采光充盈于挑高3 m的宽敞空间中，木制地板呈现居家温暖的氛围。开放式的空间格局，借由隔间拉门设计可自由划分出睡房、卫浴等私人空间，呈现舒适安心的住房感。室内设计不仅呼应饭店整体新欧风，更在空间功能与设计细部融合中日元素。寝室空间以樱花的概念设计、欧式图腾和艺术装置，流畅地演绎当代低调奢华的美学设计创意。

Interior

The Okura Prestige Taipei is designed by Chhada, Siembieda & Associates Ltd., featured in contemporary design; the rooms exude soft lighting to illuminate the spacious 3-meter-high ceiling space and accentuate the room's décor. The open space arrangements with sliding partitions allow guests to personalize their own space. Reflecting the European theme, adding both Chinese and Japanese theme in the design of its space, the bedroom is designed by the concept of cherry flower, European totem and sawing arts, expressing the creation of contemporary living style.

酒店配套

■ 餐厅

大仓久和大饭店拥有三间富有魅力的特色餐厅，包括中式、日式、自助餐厅与一间酒吧。源自东京大仓，深获世界高度评价的山里日本料理，由来自东京大仓的资深主厨亲自主理传统会席料理，引领宾客探究精细无比又余韵无穷的正统日本料理。同样为东京大仓特色招牌餐厅、以粤菜为精髓的桃花林中华料理，为世界各国的贵宾献上顶级精致餐厅飨宴。饭店内还设有欧风馆自助餐厅，为国内外宾客提供全天候的荟萃世界各国风味的精致佳肴。饭店内并设有The Nine烘焙坊，供应法日式烘焙面包与精品蛋糕。位于一楼大厅的珍珠酒吧，以时髦与精妙的设计格调，呈现现代沉稳风格，为品味各国佳酿的自在空间。

Services and Amenities

■ Dining

The Okura Prestige Taipei offers three exclusive restaurants including Western and Eastern restaurants, two full bars, one in the lobby and the other next to the rooftop swimming pool. The highly esteemed YAMAZATO fine Japanese cuisine from Hotel Okura, YAMAZATO at Hotel Okura Amsterdam has been awarded 1 star in a Michelin Guidebook will bring the traditional Kaiseki Cuisine and Tempura. The Teppanyaki area allows guests to enjoy the flavors of fine meats and seafood cooked in another traditional way. The TOH-KA-LIN is a reputable Chinese restaurant from Hotel Okura which will serve guests around the world a traditional Cantonese cuisine. Not only does the Okura satisfy the Eastern palate, but also has a Continental Room buffet restaurant which serves an all-day buffet featuring food from around the world. There is also a bakery shop that provides exclusive gift sets on the first floor which provide options for the busy travelers. The comfortable design of the The Pearl lounge & bar in the lobby provides a respite for the guest for a savor drink while the rooftop pool bar allows the guest to enjoy drinks against the surroundings of Taipei City.

■ 宴会与会议

位于饭店 3 楼的宴会厅，拥有 108 m^2、挑高 4.5 m 的空间，最多可容纳 24 桌中式宴席，亦可弹性调整、区隔为多间宴会与会议空间。

同样位于 3 楼的桃花林中华料理餐厅以新派中国设计为主要风格，宴会厅的设计即以中式风格为设计延伸，在挑高的落地窗与大门上镶着牡丹花元素，加上西式水晶灯点缀，充盈着当代经典设计的风格，配备先进的灯光影音设备，呈现功能与美学之间的平衡，展现日本追求自然和谐的概念文化特色。

■ Meetings and Events

The Okura Prestige Taipei offers functional space with mobile partitions to customize any special events and professional meetings. The multi-purpose and spacious exemplifies the combination of classic and modern design. Equipped with advanced lighting, audio and video facilities, the banquet hall demonstrate a balance between function and aesthetics, highlighting the epitome of Japanese culture.

Toh-Ka-Lin Chinese Restaurant is also located on the 3rd floor as the main style to the new breed of Chinese design, the design of the banquet hall follows the Chinese style design extension of the ceiling French windows and door trimmed with the peony as the element, together with the Western Crystal Light embellishment, filling the style of contemporary and classic design, equipped with advanced audio and video lighting equipment, showing a balance between function and aesthetics, and to show Japan's pursuit of the natural harmony and cultural characteristics.

■ 休闲设施

人与水密切相关，水为生命之源。大仓健身俱乐部以 Aqua 水感为整体概念，拥有宽敞的顶级舒压水疗空间与顶楼露天温水游泳池，让宾客以水滋润心灵，放松精神，让身心与生活恢复平衡和谐。

■ Relaxation Facilities

Relationship between human being and water is closely linked, for it is the fountain of life, themed in "Aqua", The Okura Health Club provides guests with a top spacious spa space and out- door pool on the roof.

■ 客房

饭店拥有208间配备先进的客房。强调日式生活机能的客房以打造私人豪宅为空间概念，包括菁英客房、大仓菁英客房、尊荣客房、行政客房、名门套房、久和套房、总统套房7款各具风格的客房。每间房间皆备有日式机能卫浴设备，宽敞的浴缸配有21寸防水电视与淋浴空间、免治马桶，可感受全然舒适无压的氛围。客房内拥有大型高画质电视、高速无线上网、电子保险箱与高级生活必备品等齐全的住房机能设备。房客同时能享受其他楼层的专业健身房、室外游泳池、其他专业舒压设备，以及大仓特有的不受任何干扰的贴心服务。客房卫浴以日式生活机能为概念，浴缸、淋浴区与卫生设备分别规划为二空间区块，让旅人的使用面更宽广，能独立使用设备。部分客房设有私人阳台，市区景观无距离进入眼帘。

■ Rooms

The hotel offers 208 exclusively designed guestrooms varied in 7 different room types, including Prestige Room, Okura Prestige Room, Premium Prestige Room, Executive Room, Junior Suite, The Suite and Royal Suite. Each room is equipped with separate dry–wet bathing facilities, one with a large bathing tub with 21 inch waterproof TV and an independent shower room and another powder room which provides a multi–functional toilet. An HD TV, high–speed internet connectivity, an electronic safe all ensure customers convenience and comfort. In addition, guests can enjoy a workout at the fitness centre, a leisurely swim in the outdoor heated swimming pool, and unwind at the massage facility, all without any disturbance. Concept in contemporary living function, separating bath tub and shower to give guests a larger and more comfortable space. Some guest rooms have private balcony that offer the exclusive city view to our guests.

The Okura Prestige
TAIPEI

The Okura Prestige
TAIPEI

The Okura Prestige
TAIPEI

The St. Regis Doha

多哈瑞吉酒店

Keywords 关键词

Personalized Service 个性化服务

Elegance 优雅

Meticulous and Considerate 细致周到

Luxury 奢华

酒店地址：卡塔尔首都多哈市多哈西湾

电　　话：+ (974) 44460401

Address: Doha West Bay, Doha, Qatar

Tel: + (974) 44460401

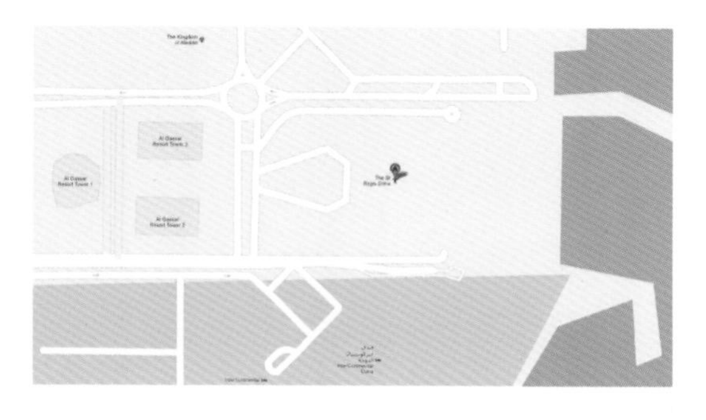

品牌链接

瑞吉酒店是喜达屋旗下的奢侈品牌，创于 1999 年。瑞吉酒店面向高端商务和休闲旅客提供完美无暇的定制服务，是世界上最高档酒店的标志。作为优雅与精致的最佳典范，瑞吉酒店及度假村对卓越的追求可谓精益求精。确保每次住宿都能达到客人的最高标准，满足其高贵典雅的独特品味。酒店还通过瑞吉迷（St. Regis Aficionado(SM)）诚邀客人利用独享特权表达对瑞吉的热爱之情，其中包括西斯廷教堂私人游、品尝彼特鲁庄园稀有精选佳酿的难得机会或有私人主厨和品酒师相伴的乡村滑雪之旅。

而今，瑞吉酒店及度假村已经遍布全球的多个精彩地点。如伦敦、纽约、新加坡、巴厘岛——每家酒店都能让宾客领略到当地的迷人风情和独特风韵。此外，每家酒店还一如既往地承袭了首家瑞吉酒店所确立的豪华与精致标准，并在不断发展中又纳入五种与众不同的设计风格：都市庄园、玻璃屋、精致半球、旅程终点和天堂乐园。每种设计风格均以独特手法完美再现了瑞吉品牌的丰富精髓与悠久传统，例如宽大的楼梯、熠熠闪光的吊灯、雅致的图书馆、琳琅满目的酒窖、标志性壁画和青铜正面装饰。

About St. Regis

As exemplars of elegance and refinement, St. Regis hotels and resorts are uncompromising in their pursuit of excellence. Every stay is commissioned to meet guests' highest standards and refined to express the subtlety of their unique tastes. Through St. Regis Aficionado(SM), guests of the hotel are invited to indulge their passions with exclusive privileges such as a private tour of the Sistine Chapel, the opportunity to sample a rare vertical selection of Chateau Petrus, or a back country ski trip with a personal chef and sommelier.

St. Regis is Starwood's main luxury brand, launched in 1999. It is the sign of the most upscale hotels in the world providing flawless and bespoke service to high-end leisure and business travelers. Today St. Regis hotels and resorts can be found across the globe. London, New York, Singapore, Bali - each is an entrance into a captivating world of seduction and a unique expression of its location. The standard of opulence and sophistication established by the original St. Regis is honored in every address, but has evolved to include five distinct design interpretations: Metropolitan Manor, Glass House, Hemispheres, Journey's End and Paradise Found. In each, the essence of the brand and its rich traditions is brought to life through signature features such as grand staircases, glittering chandeliers, handsome libraries, vast wine vaults, iconic murals and bronze façades.

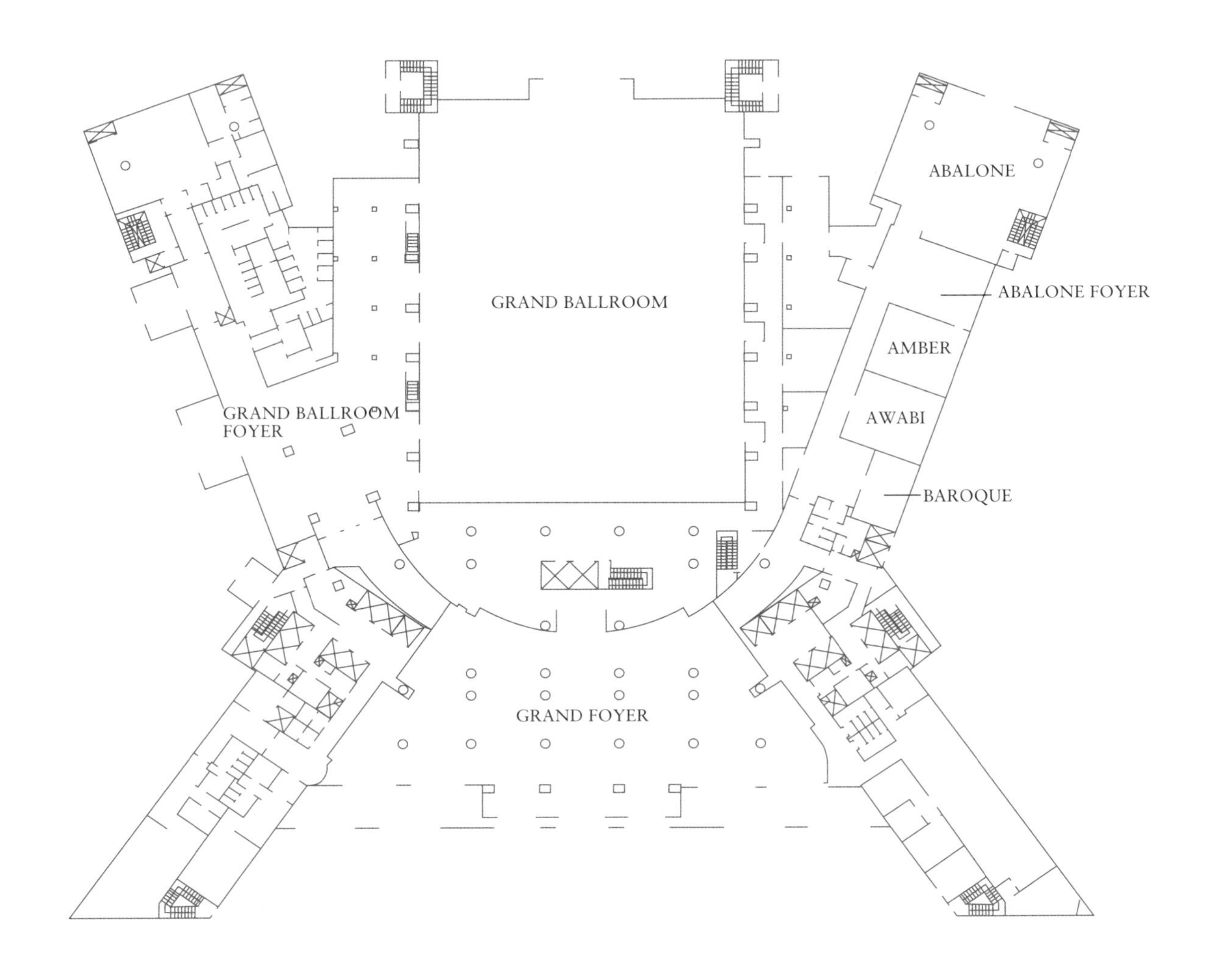
ABALONE
ABALONE FOYER
GRAND BALLROOM
AMBER
AWABI
GRAND BALLROOM FOYER
BAROQUE
GRAND FOYER

酒店概况

多哈瑞吉酒店是多哈五星级豪华酒店，拥有 336 间客房，其中包括 70 间经典套房。这座多哈经典地标性建筑的灵感来源于其周边的沙丘和古建筑。另外，酒店还拥有 22 间私人护理室的招牌 Remède 水疗中心，让客人逃离烦恼，精力无限充沛。

Overview

The St. Regis Doha is a five star luxury hotel in Doha with 336 guest rooms including 70 seductive suites. The surrounding sand dunes and ancient architecture were the inspiration for this timeless Doha landmark. Escape to a sanctuary of rejuvenation at the signature Remède Spa featuring 22 private treatment rooms.

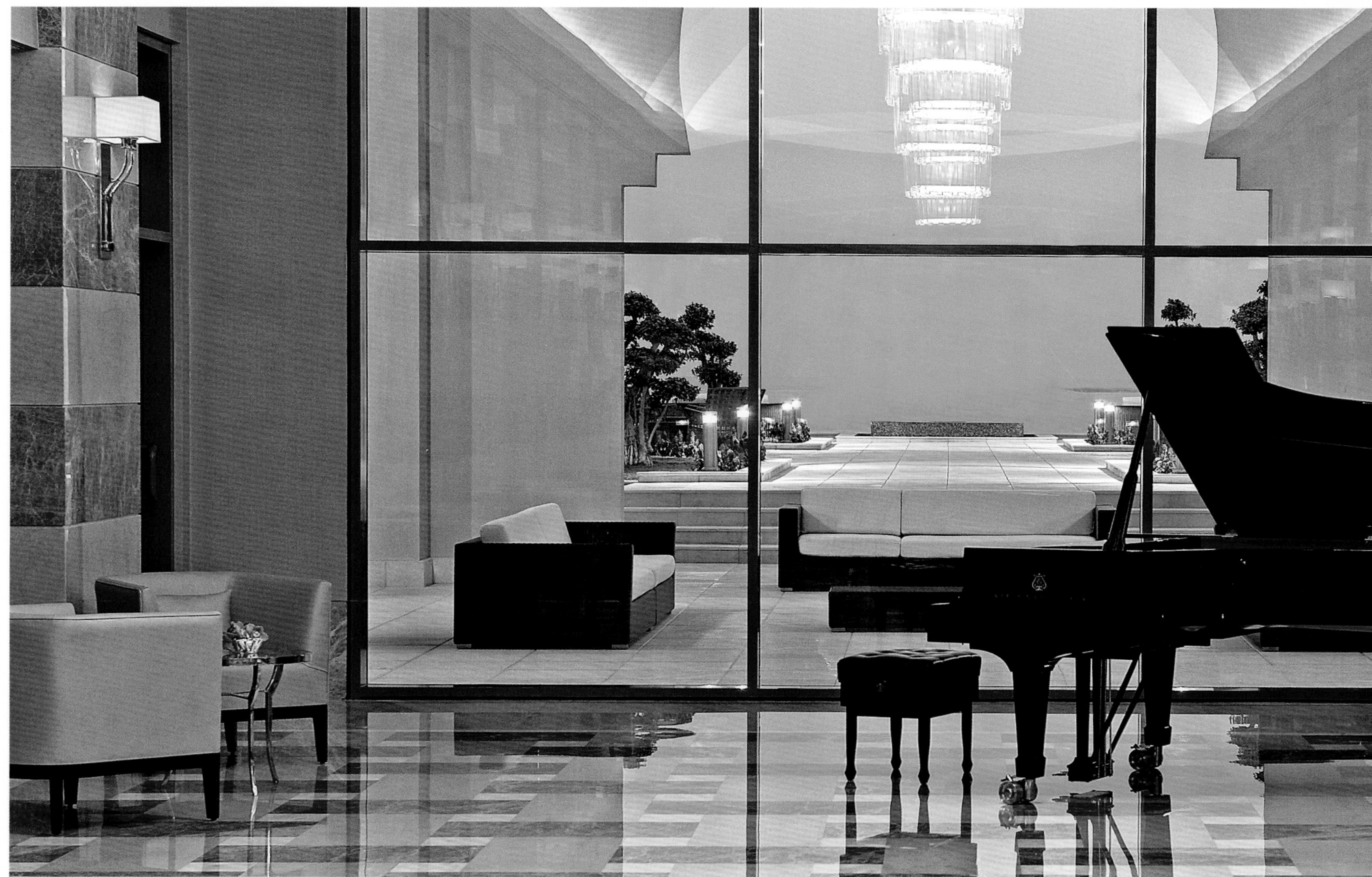

酒店特色

多哈瑞吉酒店为每位客人提供全程细致周到的私人管家服务，客人未入住酒店前就可以感受到此种优质服务，并承诺为所有客人营造珍藏一生的入住体验。

Feature

Meticulous and discreet personal service is flawlessly delivered even from preceding visits. The St. Regis butler and a dedicated staff bring virtually every request and whim to brilliant fruition.

酒店配套

■ 庆典及会议

酒店拥有超过 4 000 m^2 的活动空间，设有 8 间会议室，包括自然采光、带俯瞰阿拉伯湾观景平台的多哈本地最大的宴会厅，是举行会面、展销、婚礼、社交、商务会议的最佳场所。

■ 精品店

瑞吉酒店为客人提供了多样可选择的旅游用品、旅游纪念品、泳装、报纸和杂志。精品店位于一楼，在商务中心附近，每日开放。

Services and Amenities

■ Celebrations and Other Events

With over 4,000 sqm of event space, eight meeting rooms including the largest ballroom with natural daylight in Doha and a terrace with extensive views over the Arabian Gulf, there is no finer address in Qatar to host a meeting, exhibition, wedding, social event or conference.

■ Boutique

The St. Regis Boutique offers guests a selection of travel supplies, souvenirs, swimwear, daily newspaper and magazines. The boutique is located on the ground floor, next to the business center and is open daily.

■ 水疗中心

瑞吉多哈酒店22间护理室中的任何一间，都可享受奢华的体验和静谧的氛围。被誉为世界最好的水疗中心之一，实现服务与品质的完美融合，正如瑞吉酒店实现卡特尔神秘风情与文化的完美融合。

■ Spa

Every single appointment at one of the twenty-two Remède Spa treatment rooms at The St. Regis Doha is a journey into a fabulous world of pampered indulgence and glorious serenity.

■ 健身

多哈瑞吉酒店提供高档健身设备，包括跑步机、椭圆机、直立单车、倾斜单车，且配有独立电视及iPod扩充口。

■ Fitness Center

Fitness facilities at The St. Regis Doha are, of course, state of the art and include treadmills, elliptical, upright and reclining bikes with individual television sets and iPod docking stations.

Westin Changbaishan Resort | 长白山万达威斯汀度假酒店

Keywords 关键词

Modern Style 现代风格

Stylish and Eclectic 低调优雅

Skiing Park 滑雪场

Rustic Atmosphere 乡土气息

酒店地址：中国吉林省长白山白云路 333 号

电　　话：+86 439 6986999

Address: NO.333 Baiyun Road, Changbaishan, Jilin, China

Tel: +86 439 6986999

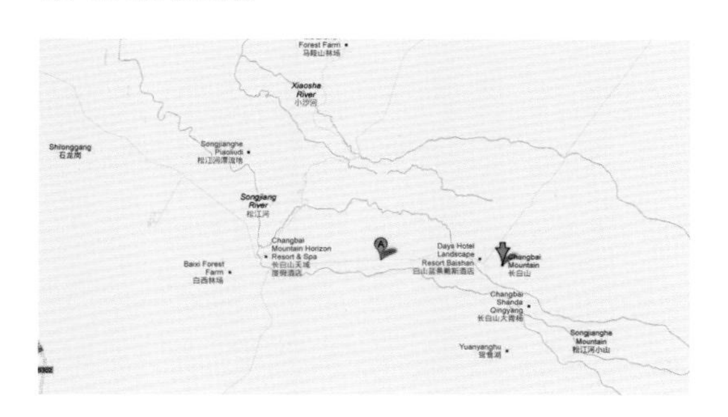

品牌链接

威斯汀酒店及度假村是喜达屋酒店与度假村国际集团的品牌之一，在全球超过 36 个国家和地区拥有 180 多家酒店和度假村。威斯汀酒店及度假村可使客人尽享康乐元素。清爽氛围、创新计划和周到设施有助于为客人提供一次乘兴而来、满意而归的绝佳住宿。现代化的设计、悉心周到的服务和令人活力焕发的氛围成为全球 180 多家威斯汀酒店及度假村的招牌。无论在西班牙享受高尔夫挥杆之乐，在巴厘岛体验惊险浮潜之游，还是在时代广场悠闲观光游览，威斯汀总能带给客人与众不同的完美体验。

About Westin Hotels and Resorts

With more than 180 hotels and resorts opened in 36 countries and areas, Westin Hotels and Resorts is a hotel brand of the Starwood Hotels and Resorts Worldwide Inc., which is the most high-end hotel company in the world. The guests indulge in elements of well-being in Westin. The refreshing ambiance, innovative programs and thoughtful amenities can provide the guests with a stay that leaves one feeling better than when arrived. More than 180 hotels and resorts worldwide are defined by modern design, instinctive service and rejuvenating atmosphere. Whether golfing in Spain, snorkeling in Bali, or sightseeing in Times Square, Westin delivers a perfect experience unlike any other.

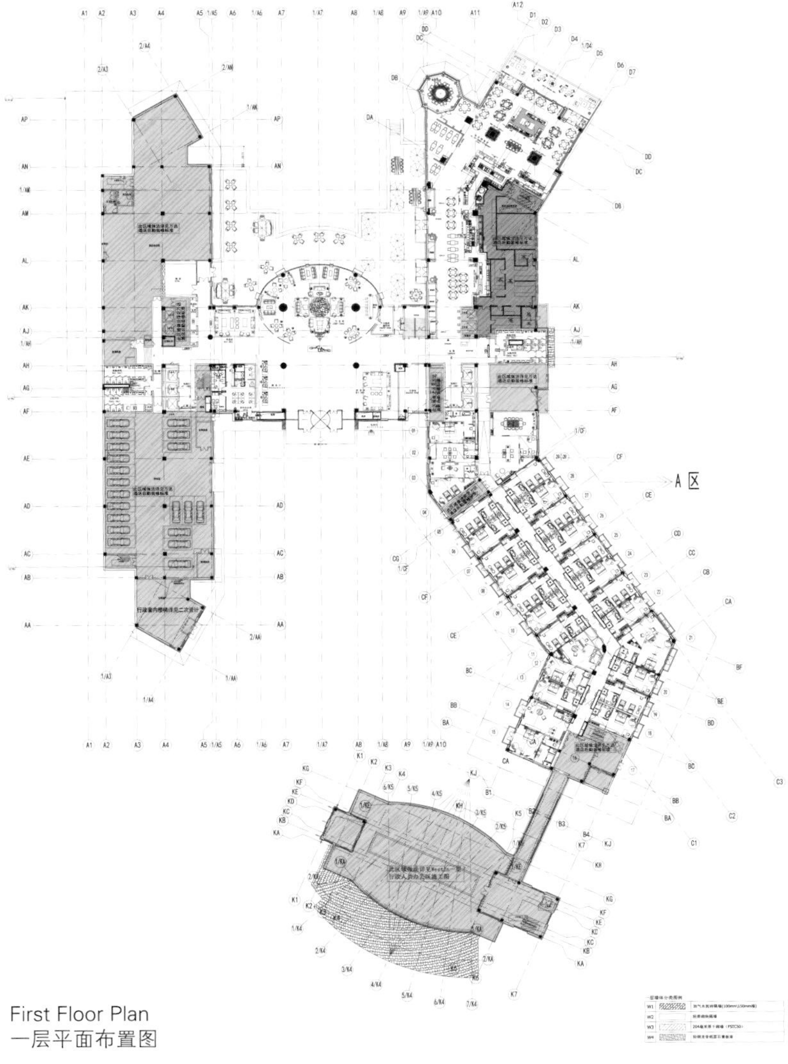

First Floor Plan
一层平面布置图

WESTIN

酒店概况

长白山万达威斯汀度假酒店位于长白山国际度假区内，距离长白山机场 20 分钟车程。酒店毗邻亚洲首屈一指的滑雪场。拥有 262 间装修典雅时尚的客房和套房，其中包括 29 间套房和 1 间 320 m^2 的总统套房。酒店的多功能空间拥有 10 间独立的会议室，总面积超过 2 400 m^2，其中包括无柱式豪华宴会厅和宽敞的迎宾区。

Overview

The Westin Changbaishan Resort is ideally located within Changbaishan International Holiday Resort, only twenty minutes drive from Changbaishan airport. The hotel is closely next to the skiing paradise of Asia, where people could experience the exciting of skiing. The Changbaishan Wanda Westin Hotel boasts a total of 262 guest rooms and suites with fashionable and elegant decoration. Among them, there are 29 suites and one presidential suite. The multi-functional space includes 10 independent conference rooms, which include column free luxury banquet hall and spacious receiving area with total area exceeding 2,400 m^2.

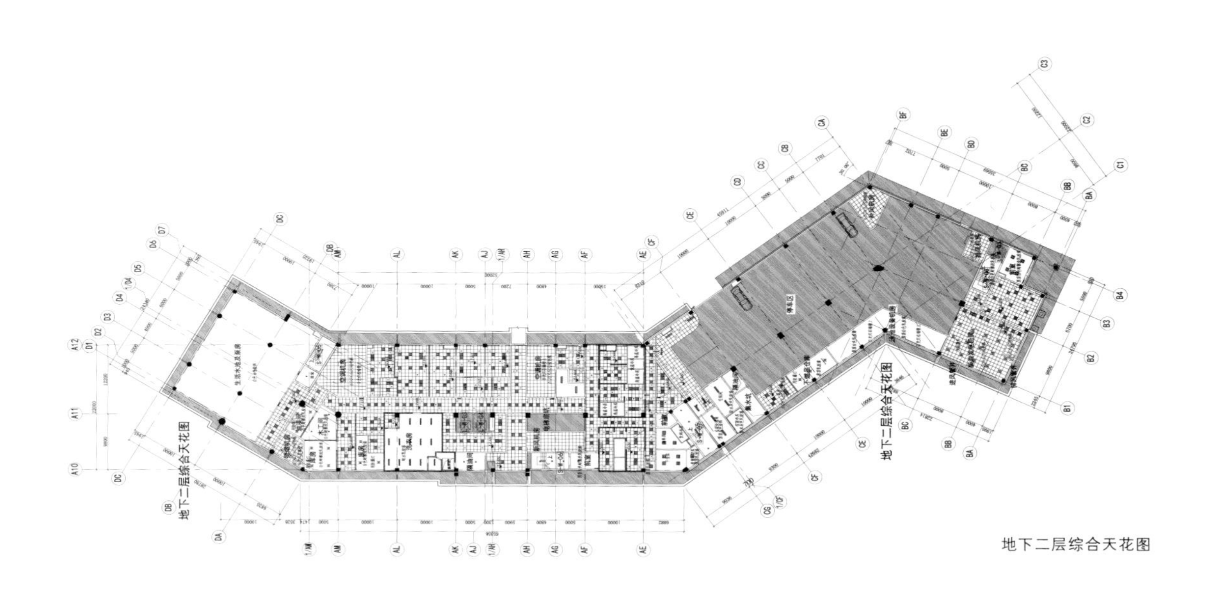

地下二层综合天花图

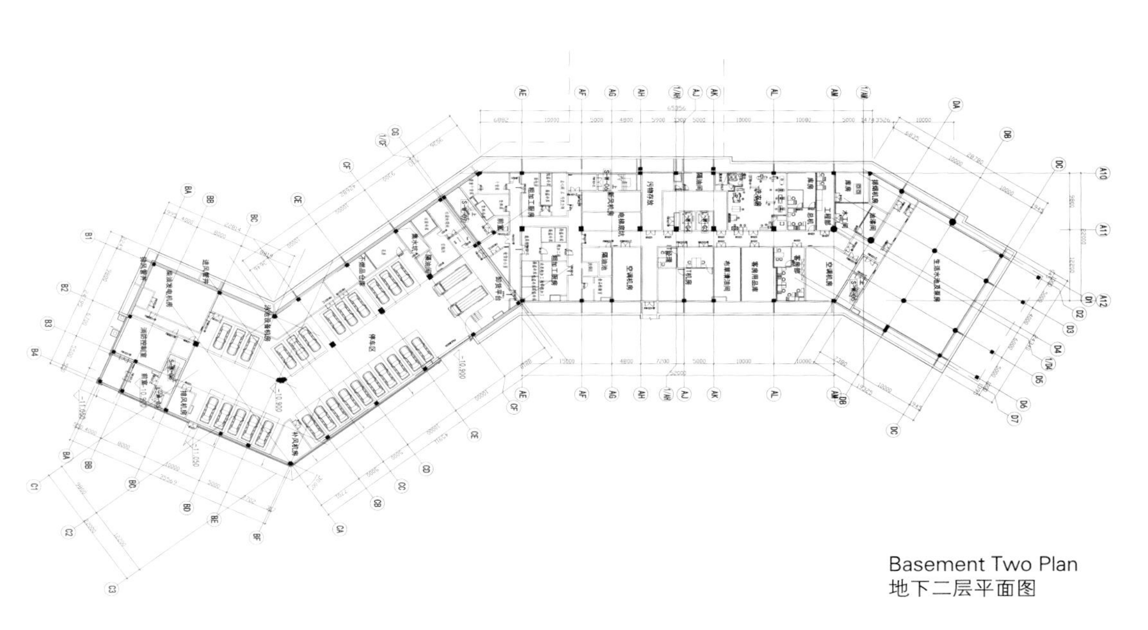

Basement Two Plan
地下二层平面图

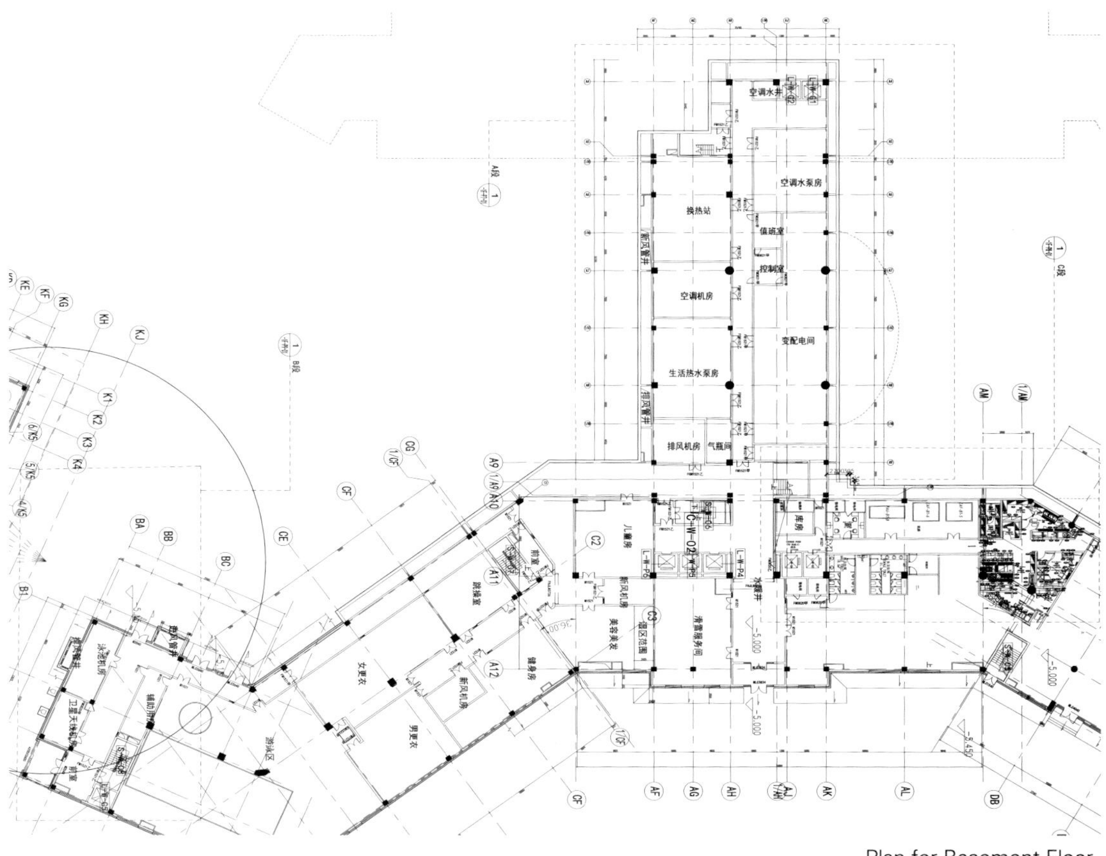

Plan for Basement Floor
地下一层总平面图

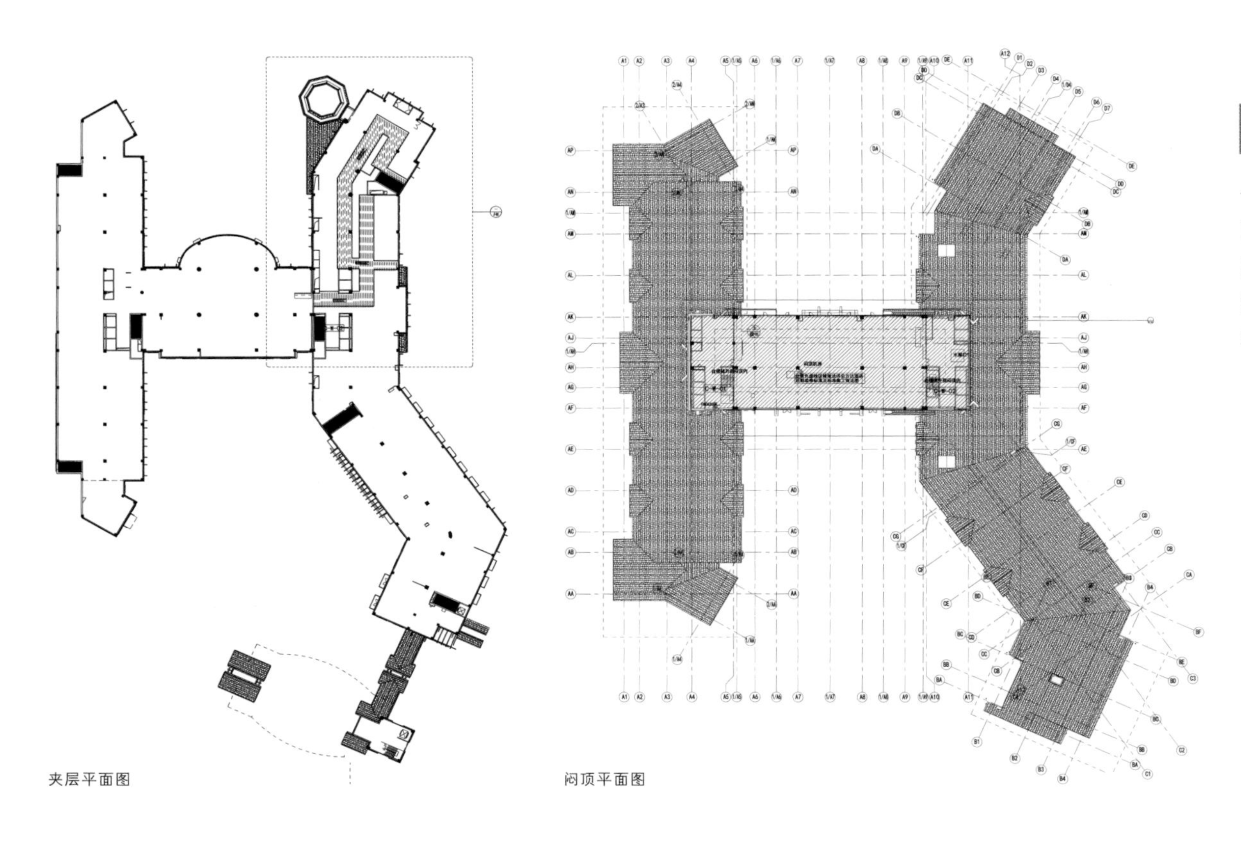

夹层平面图　　闷顶平面图

酒店特色

长白山万达威斯汀酒店坐落于万达长白山国际度假区的中心地带，酒店毗邻亚洲首屈一指的滑雪场，不仅可以使游客体验到滑雪的激情，温暖的季节亦可以登山远足，呼吸清新空气，全年均适合观光游览。

Feature

The Westin Changbaishan Resort is within the central area of Chang-baishan Intertainal Resort, which is not only among the first class ski-ing field for tourists to experience the passion of skiing, but also a populous resort to climb on top of mountains and breathe fresh air in warm season.

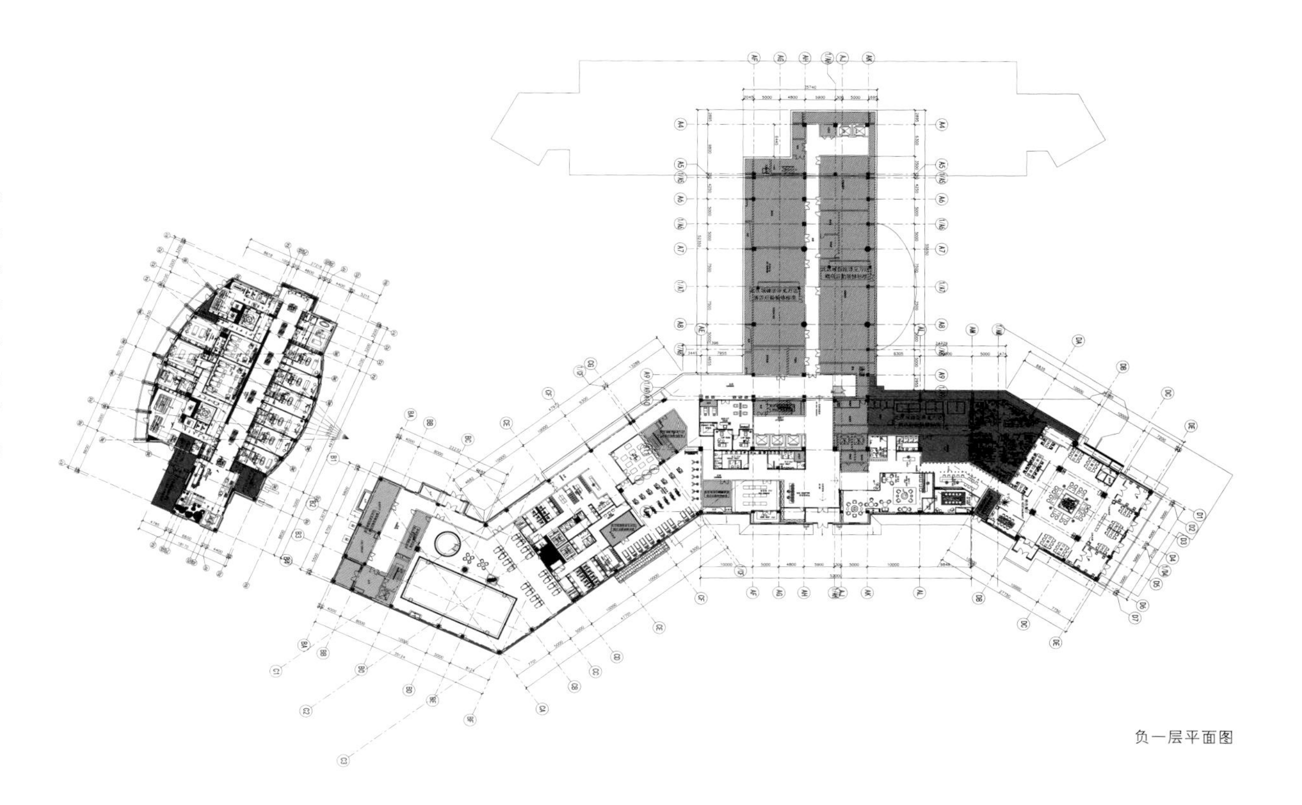

负一层平面图

酒店室内

酒店客房内均采用现代设计风格的元素和中式乡土气息浓厚的低调的自然色。深色方格形的木地板、灯笼样的灯饰以及长毛绒的小地毯为客人休息营造了良好的氛围。透过落地窗户，客人还可以欣赏到周围滑雪场的风光，有些房间还有观景阳台供客人呼吸新鲜空气。

Interior

The hotel rooms beckon with a balance of modern design elements and rustic Chinese style in subdued natural tones. Lattice, dark wood floors, lantern-style lamps, and a plush area rug create an ideal atmosphere for rest and recentering. Floor-to-ceiling views of the nearby ski slopes inspire, and some rooms offer balconies for fresh air.

酒店配套

■ 大堂吧

温馨的大堂吧是适宜畅谈和休息的绝佳场所。透过落地窗可以欣赏到附近如画般的雪山美景。鸡尾酒、热茶、小食，坐在沙发和扶手椅上，客人可以放松身心，尽情享受这个低调优雅的空间。

■ Maru 韩国餐厅

现代风格的 Maru 韩国餐厅为客人提供意想不到的、精心设计的、正宗的韩国口味菜肴。充满艺术感的餐厅装饰搭配豪华座椅、浅色木雕，呈现出高贵典雅的氛围。餐厅配备的室外就餐区域可供夏季使用，可以在品尝美味佳肴的同时呼吸新鲜空气，欣赏到周围美丽的山景。

■ 知味餐厅

低调朴素、现代风格的知味餐厅充满闲适的格调，开放式的厨房与客人之间形成良好的互动氛围。早餐可以吃到营养诱人的苹果派甜甜圈搭配蜂蜜，午餐或晚餐有菠菜沙拉、烤南瓜籽配低脂酸奶酱、豆腐汉堡配日式烧酱以及烤鲑鱼等等，丰富多样的菜肴供客人随心选择。

Services and Amenities

■ Lobby Lounge

Inviting and cozy, the Lobby Lounge is just the place to savor conversation or simply relax next to floor-to-ceiling windows revealing picturesque hillside views. From cocktails and hot tea to light snacks, there is plenty to soothe you in this circular space filled with a stylish, eclectic mix of sofas and armchairs under a high domed ceiling.

■ Maru Restaurant

Stylish and distinctive Maru is an unmatched destination for unexpected flavors found in creatively prepared, authentic Korean fare. Artful, refined décor comprises a mix of plush seating, light woods, and a natural palette, all centered on a dynamic light wood sculpture. An outdoor dining area is open during the warmer months, for fresh air and mountain views.

■ Seasonal Tastes

Stylish, understated décor and modern accents set a relaxed tone at Seasonal Tastes, while an open kitchen creates an interactive atmosphere. Enjoy a tempting, nutrient-rich choice such as an apple-filled donut with honey at breakfast. At lunch or dinner, benefit from the nutritious synergy found in our delicious spinach salad with roasted pumpkin seeds and low-fat yogurt dressing; a tofu burger with teriyaki sauce; or our roast salmon with braised Romaine lettuce and chorizo.

■ 威斯汀健身馆

在健身馆里，客人可以使用最先进科学的健身器材，如跑步机、静止脚踏车、踏步机等。此外，健身馆还提供瑜伽课程和个人训练指导。

■ Westin Workout Gym

The gym enhances clients' well-being with an array of state-of-the-art equipment such as treadmills, stationary bikes, step machines, and more. The gym also offers yoga classes and personal training.

■ 威斯汀天堂 Spa

天堂 Spa 里环境优雅，技师技艺精湛，独特的疗法与传统的中药艺术相结合，通过现代手段达到身心的完全放松。绿植、水景、柔和的灯光以及美妙的音乐帮助客人驱赶各种压力。

■ 威斯汀儿童俱乐部

威斯汀儿童俱乐部是 3~12 岁儿童的冒险乐园，这里有专门的员工进行看护。彩色的空间吸引着孩子们跃入其中玩滑梯、过家家，大一点的孩子可以玩滑板、折纸，看电影等。

■ 商务中心

商务中心提供最先进的设备，可以协助客人在旅途中完成工作。专门的服务人员将负责协助客人处理有关问题。商务中心提供的大型会议室还可以欣赏到庭院的景色，会议室内配有 LCD 投影仪以及 8 人的会议桌。

■ Westin Heavenly Spa

Westin Heavenly Spa– is an exquisite environment where unique treatments marry the traditional art of Chinese medicine with a contemporary approach to holistic well–being, for total relaxation. Greenery, water features, soft lighting, and music accent the sleek surroundings and dissolve stress.

■ Westin Kids Club

Westin Kids Club– is an adventure–filled heaven designed for children from three to 12 years old. Supervised by enthusiastic staff, the colorful space inspires kids to jump into the action with toddlers' toys like a slide and toy house, plus plenty for older kids such as board games, origami, movies, and more.

■ Business Center

Staffed Business Center– is fully equipped with the latest facilities so that guests can handle tasks seamlessly during their stay. Friendly associates are available to answer questions or provide assistance. The Business Center also offers a meeting room with views of the courtyard, an LCD projector and screen, as well as a conference table for eight.

Sheraton Qingyuan Lion Lake Resort | 清远狮子湖喜来登度假酒店

Keywords 关键词

Arabian Expressions 阿拉伯风情

Lakeside Vacation 临湖度假

Scarce Resources 稀缺资源

Feast for the eyes 赏心悦目

酒店地址：中国广东省清远市清城区横荷街狮子湖大道 1 号

电　　话：0763-8888888

Address: NO.1 Lion Lake Avenue, Henghe Street, Qingcheng District, Qingyuan, Guangdong, China

Tel: 0763-8888888

品牌链接

喜来登酒店与度假村集团（Sheraton Hotels and Resorts）是喜达屋（Starwood）酒店集团中最大的连锁旅馆品牌，而它也是集团中第二老的酒店品牌（最老牌的是威斯汀）。喜来登的酒店型态有许多种，从一般的商业旅馆到大型度假村都有；喜来登品牌一直力图维持高品质形象，在世界上的喜来登酒店有超过一半被当地机关评选为五星级酒店。喜来登酒店据点分布极广，遍布五大洲，从香港到斯里兰卡到埃及及津巴布韦等国都可见其旅馆。喜来登总部在美国纽约的白原市。

喜达屋酒店都有良好的选址，主要分布在大城市和度假区。集团酒店选址的标准是：所在区域的发展史表明，该地区对提供全方位服务的豪华高档酒店有大量、持续增长的需求。作为酒店业豪华高档细分市场中最大的酒店集团，喜达屋酒店的规模有力地支持它的核心市场营销和预定系统。喜达屋酒店在把重点放在豪华高档细分市场同时，其各种品牌分别侧重于该市场中不同的二级市场。喜达屋酒店在赌场业也占据着重要的位置，它主要是通过 Caesars 品牌来经营此业务。喜达屋酒店为休闲度假旅游者提供着宾至如归（home-away-from-home）的服务。

About Sheraton Hotels and Resorts

Sheraton Hotels and Resorts is Starwood Hotels and Resorts Worldwide's largest and second oldest brand (Westin being the oldest). It occupies many types of hotel properties, ranging from general commercial hotels to large-scale resorts with high-quality image; more than half of Sheraton hotel worldwide are recognized as five-star hotels, with a widespread distribution across five continents i.e. Hong Kong, Sri Lanka, Zimbabwe etc. It headquarters in White Plains, New York.

Starwood hotels are located mainly in big cities or resort districts, in accordance with the site selection standards of group hotel that the selected districts own an increasing and great demand for full-serviced luxurious high-end hotels. As the greatest hotel group in the luxurious high-end market of hotel industry, Starwood holds its coral market promotion and booking system supported by its scale. It lays its emphasis on luxurious high-end market and its brands are targeted at the secondary markets. It also has a great achievement in casino industry, mainly represented by Caesars brand. Sheraton hotels address themselves to offer the visitors services of home-away-from-home.

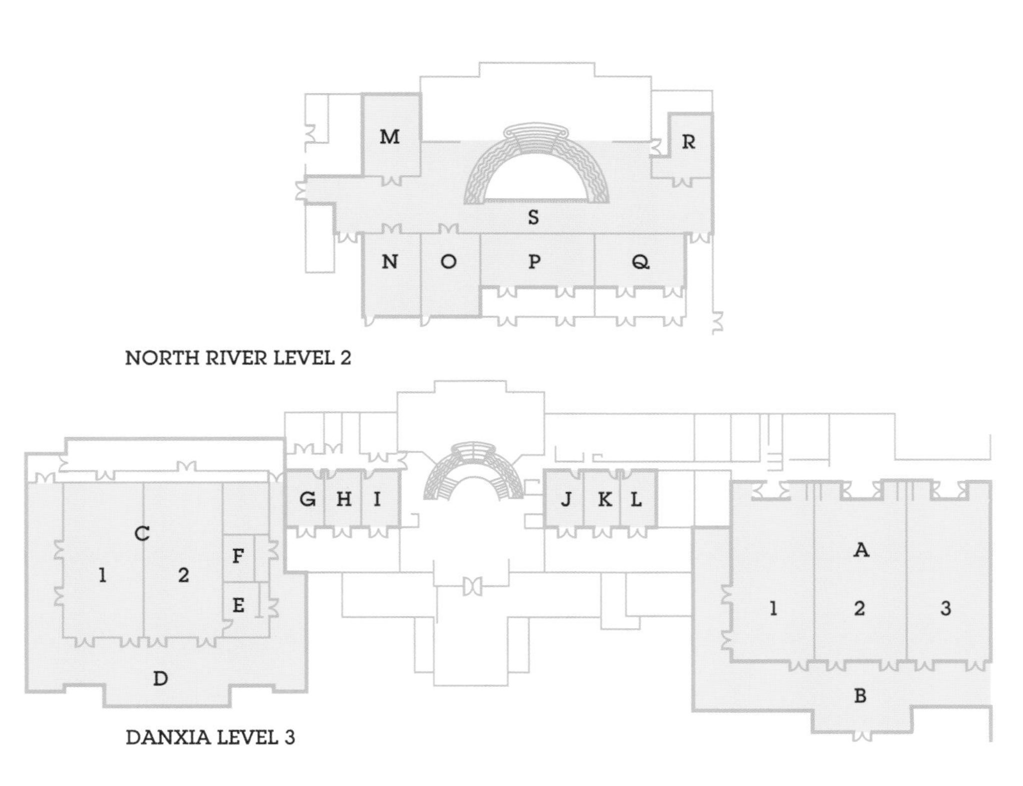

KEY

A THE GRAND BALLROOM
GRAND I
GRAND II
GRAND III
GRAND I+II/II+III

B PRE-BALLROOM AREA
C SHERATON BALLROOM
SHERATON BALLROOM I
SHERATON BALLROOM II
D PRE-BALLROOM AREA

E VIP ROOM
F BOARD ROOM
G DANXIA I
H DANXIA II
I DANXIA III

J DANXIA IV
K DANXIA V
L DANXIA VI
M NORTH RIVER I
N NORTH RIVER II

O NORTH RIVER III
P NORTH RIVER IV
Q NORTH RIVER V
R NORTH RIVER VI
S PRE-FUNCTION AREA

酒店概况

清远狮子湖喜来登度假酒店坐落于广东省清远市横荷街狮子湖大道 1 号。狮子湖喜来登酒店是具有浓郁阿拉伯风情建筑的临湖度假酒店。酒店内附设购物商场、商品街、水幕电影、园林景观。酒店拥有完善的喜来登配套设施，尽显喜来登特色。

Overview

Sheraton Qingyuan Lion Lake Resort is located in NO.1 Lion Lake Street, Henghe Road, Qingyuan, Guangdong, with intense Arabian temperament. There equips with shopping mall, commercial street, water screen film, garden landscape. The classic Sheraton facilities are also provided in the hotel to demonstrate Sheraton characteristic.

酒店特色

酒店主打特色阿拉伯建筑，以空中花园的建筑形态、一千零一夜人物传说雕塑及壁画打造一个具有异域风情的商务会议度假型酒店。

Feature

The hotel features Arabian architectural style, with architecture forms of garden in the air, sculptures of legendary figures in 1001 nights and paintings to create a completely alien business and vacation hotel.

酒店外观

穹顶是阿拉伯建筑的特色标志，一般的建筑由一个大穹顶和小穹顶组成，但是狮子湖喜来登拥有 14 个穹顶，其中最大的穹顶高达 28 m，将阿拉伯特色发挥到了极致。这些设计和建造，别具风格，在整个珠三角独树一帜，而且与整个项目的规划互相依存，互为补充的，将居者的感受也详细地考虑进去。酒店是狮子湖的重要配套组成部分，同时也是狮子湖居住者的重要风景线。

Exterior

Dome is the unique feature of Arabian architecture. Generally, a big dome and a small dome consist the rooftop of ordinary buildings, but Sheraton Qingyuan Lion Lake Resort has 14 domes, with the largest rising 28 meters high as the extreme of Arabian characteristic. The distinctive architectural design and construction develops its own style in the entire Pearl River Delta. The hotel is in consistent with and complementary to the overall planning of the area, which also takes in consideration of residents' feelings. The Sheraton Hotel is a key part of Lion Lake and an important landscape for the residents of Lion Lake.

酒 店 室 内

清远狮子湖喜来登度假酒店是阿拉伯建筑艺术的体现，也是阿拉伯文化与中国文化相互融合的结晶。装饰风格上，选择了具有阿拉伯特色的饰品、地毯、艺术品摆件、灯具、穹顶、桃形拱门、拱窗及金属花格等。客房门口闪烁的贝壳来自遥远的越南和印度尼西亚，焕发出纯正异域的光彩。从东方文明走进阿拉伯世界，酒店是阿拉伯文化艺术与清远稀缺资源的完美结合，宾客身在中国就能轻松体验阿拉伯的文化魅力，同时也为传承世界文化瑰宝添砖加瓦。

Interior

Sheraton Qingyuan Lion Lake Resort reflects the art of Arabian architecture as well as the fine result of the combination of Arabian culture and Chinese culture. In decoration style, the projects chooses Arabian characteristic accessories, carpet, artistic ornaments, lights, dome, peach shape archway, arch window and metallic lattice, etc. The shells mounted on each guest room door come from very far countries like Vietnam and Indonesia, shining pure exotic color. While guests walking from Eastern civilization entering Arabian civilization, this Sheraton hotel is a perfect combination of Arabian art and culture with Qingyuan scarce resources. Guests could experience the distinctive charms of Arabian culture while contributing a small part for the inheriting of world culture.

酒店配套

■ 会议与活动

酒店拥有面积超过 20 000 m^2 的宽敞活动空间，包括 16 间赏心悦目的多功能厅，有些可以饱览湖泊或花园的迷人美景。豪华的狮子湖宴会厅装潢精美，并拥有高达 7 m 的天花板，宴会厅内可轻松容纳 1 600 位宾客，并可举办鸡尾酒会，还可以灵活划分成三间小厅。

Services and Amenities

■ Conference Hall

The hotel possesses spacious activity room with over 20,000 m^2, including 16 enjoyable multi-function-al halls, some providing charming views of the lake and garden. The decoration style of luxury Lion Lake banquet hall is exquisite and bright with 7 m ceiling, which could hold up to 1600 guests for a cocktail party and can be divided into three smaller halls as well.

■ 餐饮酒吧

酒店拥有 4 间融汇中西的餐厅及酒吧，带给宾客无与伦比的餐饮体验，提供全天候点菜及分时段自助餐的盛宴。全日制餐厅，供应当地地道口味及正宗粤菜的“采悦轩”中餐厅，最正宗最地道的班妮意大利餐厅和四面环水的卡萨布兰卡主题餐厅，均为客人提供理想惬意的用餐体验。环境宜人的大堂吧是宾客与商务伙伴或是同行旅伴交谈、品尝美味鸡尾酒的休闲放松之地，同时可欣赏波光粼粼的狮子湖；红酒雪茄吧也为红酒及雪茄爱好者提供沟通交流的好去处。

■ Restaurants and bars

There are four restaurants and bars in total mixing both Chinese and Western Style and providing fantastic dining experience to guests. 24 hours ordering and periods of buffets are available in the Banquet Restaurant; Caiyuexuan Chinese Restaurant services local flavor dishes and orthodox Cantonese dishes; the authentic Bennett Italian Restaurant and water-circling Casablanca Theme Restaurant all provide pleasant and agreeable dining experience for guests. The enjoyable Lobby Bar offers room for partners or friends to chat, taste cocktails and relax while enjoying the sparkling Lion Lake. The Red Cigar Bar provides a good place for cigar-lovers to taste red wines and communicate with friends.

■ 娱乐设施

酒店配有一个室外泳池，一个室内恒温泳池，各带一个儿童池，可供家庭尽享亲子之乐；高尔夫爱好者可在狮子湖国际高尔夫度假俱乐部的场地上潇洒挥杆；狮子湖内亚洲顶级音乐水幕呈现360度立体环绕水帘电影。

■ Entertainment Facilities

The hotel is equipped with an outdoor pool, an indoor pool with constant temperature, each with a children's pool for families to enjoy happy time. Golf-lovers could swing their clubs on the ground of Lion Lake International Golf Vacation Club. The Asian top class music water curtain could entertain guests for a 360 degree three-dimensional water screen movie.

Sheraton Macao Hotel, Cotai Central | 澳门喜来登金沙城中心酒店

Keywords 关键词

Gracefully Decorated 装潢雅致

Leisure 休闲度假

Large-Scale 规模大

品牌链接

喜来登酒店与度假村集团（Sheraton Hotels and Resorts）是喜达屋（Starwood）酒店集团中最大的连锁旅馆品牌，而它也是集团中第二老的酒店品牌（最老牌的是威斯汀）。今日的喜来登品牌是在1937年出现的，当时两位企业家Ernest Henderson以及Robert Moore在马萨诸塞州的斯普林菲尔德成立了第一家喜来登酒店。1945年喜来登成为第一家在纽约证券交易所挂牌上市的连锁酒店集团。1995年福朋喜来登品牌成立，喜来登希望以合理的价格提供全方位的服务；当时很多规模较小的喜来登酒店都被改名为福朋喜来登。1998年喜达屋集团以高于希尔顿的出价收购了喜来登品牌。在喜达屋的管理领导下，喜来登开始创建更多的酒店以扩大其品牌影响力。

About Sheraton

Sheraton Hotels and Resorts is the largest hotel chain brand of the Starwood Hotels and Resorts Worldwide Inc. and second oldest brand (Westin being the oldest).The origins of the sheraton dated back to 1937 when Ernest Henderson and Robert Moore acquired their first hotel in Springfield, Massachusetts. In 1945, it was the first hotel chain to be listed on the New York Stock Exchange. In 1995, Sheraton introduced a new, mid-scale hotel brand Four Points instead of Sheraton Hotels, Which provides a comprehensive service at a reasonable price. In 1998, Starwood Hotels & Resorts Worldwide, Inc. acquired Sheraton, outbidding Hilton. Under Starwood's leadership, Sheraton has begun renovating many existing hotels and expanding the brand's footprint.

酒店地址：中国澳门氹仔岛路氹金光大道
电　　话：(853) 2880 2000

Address: Cotai Strip Taipa, Macau
Tel: (853) 2880 2000

Sheraton

LEVEL 4 - CASPIAN / PAMIRS / TIAN SHAN

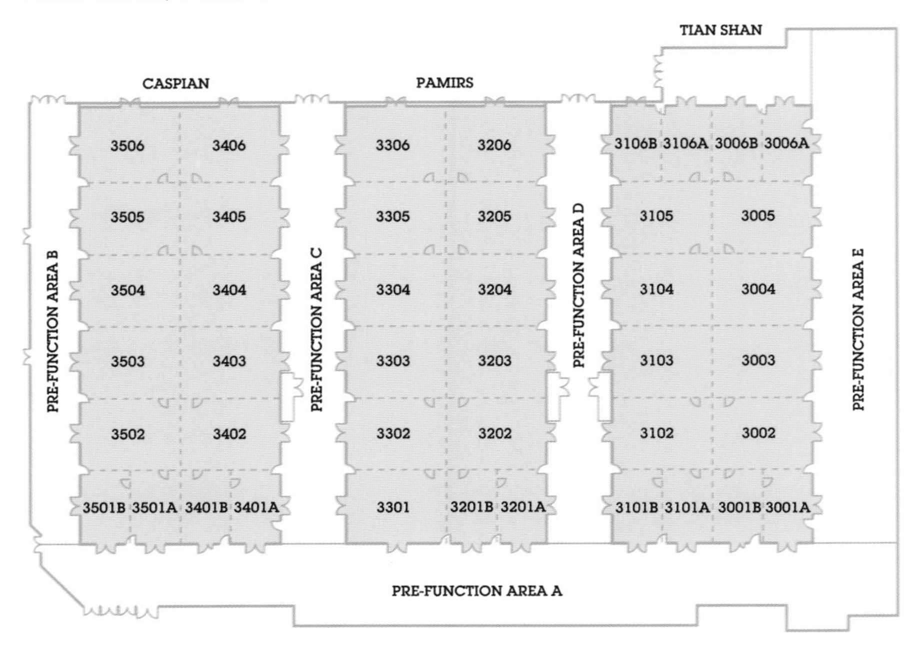

LEVEL 5 - HAMADAN

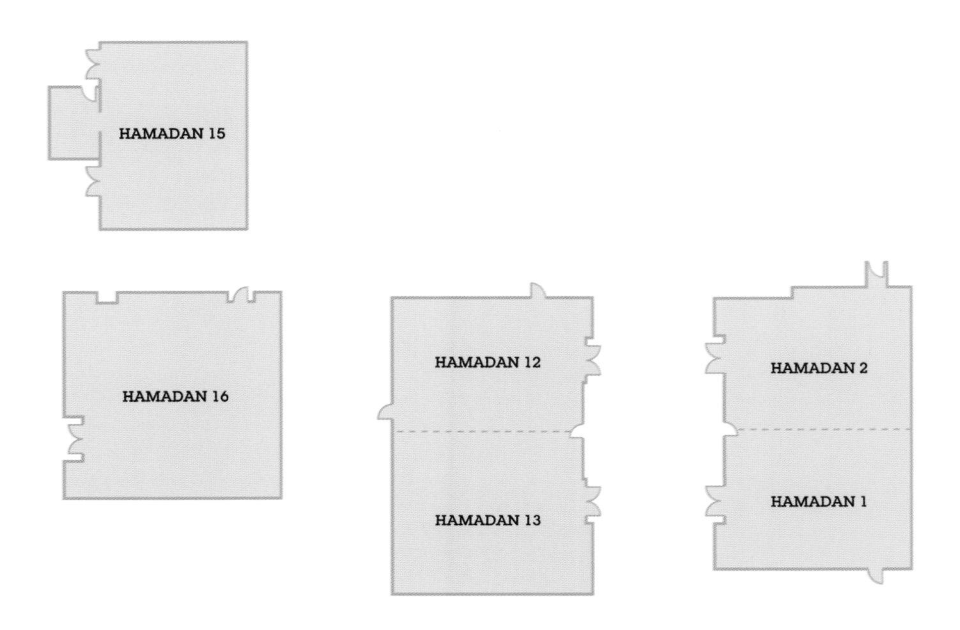

LEVEL 5 - KASHGAR

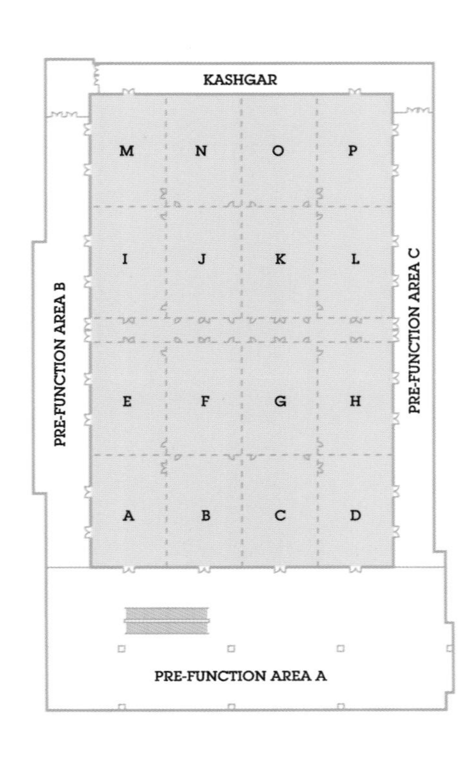

酒店概况

澳门喜来登金沙城中心酒店地理位置优越，坐落于金沙城中心。酒店设有 3 896 间装潢雅致的客房及套房，提供别具特色的社交空间，配备网络服务设施，以及会议室和接待处的喜来登行政酒廊。该酒店共有两座酒店大楼——“浩天楼”及“宏地楼”，内设 3 间各具特色的餐厅、1 间大堂酒廊、多功能的会议场地、喜来登炫逸水疗、3 个室外游泳池，以及与世界知名健身机构 Core Performance 联手设计的喜来登健身计划。

Overview

Sheraton Macao Hotel with 3,896 decorative guest rooms and suites is offering its customers unique and distinctive social space, with supporting facilities like internet access, Sheraton Executive Bar of 274 seats, conference rooms and reception center. It has two hotel buildings: Haotian Building and Hongdi Buildings, with 3 restaurants, 1 lobby bar, multi-functional conference site, Sheraton Xuanyi Spa, 3 outdoor swimming pools, and the Sheraton Fitness Plan which is co-designed with the world well-known fitness institution Core Performance.

酒 店 特 色

酒店拥有许多休闲娱乐设施，包括带凉亭的三个户外泳池、购物中心、健身中心、由 Comfyland Experience™ 规划的幼童娱乐场地和喜来登炫逸水疗中心。水疗中心设有15间护理室，为客人奉上灵感源自风水的五大元素——金、木、水、火、土的特色护理。

Feature

Ideally located within Sands® Cotai Central, the Sheraton Macao Hotel, Cotai Central provides an enticing blend of recreation and entertainment, including three outdoor pools with cabanas, a shopping mall, Fitness Center, Kid's Zone with toddler play area by Comfyland Experience™ and Shine Spa for Sheraton, featuring 15 treatment rooms and signature treatments inspired by the five elements of Feng Shui: wood, fire, earth, metal and water.

spg.
Starwood
Preferred
Guest

酒店配套

■ 餐饮

在“鲜”餐厅可品尝路氹金光大道最新鲜的海鲜料理。“班妮”餐厅供应意大利家常菜肴，而“盛宴”餐厅奉上当地美食与各种备受儿童青睐的儿童菜式。“喜柏”大堂酒廊是享受休闲下午茶的理想之地，而酒店的三间池畔咖啡厅与酒吧则是顾客啜饮鸡尾酒的最佳场所。澳籍主厨景大卫（David King）是酒店餐饮部总监。他对烹调充满热情，并凭借精湛的厨艺屡获殊荣。在他的带领下，酒店的每间特色餐厅与池畔咖啡厅均独具特色，是体验超凡餐饮体验的绝佳之地。

Services and Amenities

■ Dining

Sample the freshest seafood on Cotai Strip at Xin. Share home-style Italian cuisine at Bene and local specialties at Feast, which features a children's menu of kids' favorites, too. Experience high tea at Palms, or sip poolside cocktails at our three poolside cafés and lounges. The vision of award-winning Director of Culinary, Australian chef David King, each of the signature restaurants and poolside cafés offers a remarkable gastronomic experience reflective of his singular passion and culinary flair.

■ 会议与活动

澳门喜来登金沙城中心酒店设有约 14 126 m^2（152 050 平方英尺）的会议空间——包括喀什大型豪华宴会厅与 6 间精致宴会厅——是举办晚宴或婚礼、私密会议或国际大会的理想之地。更有专业的服务团队随时为宾客定制完美无瑕的活动套餐，并配有最先进的视听技术设备。

■ Meetings & Events

Host an inspiring conference or event at the Sheraton Macao Hotel, Cotai Central, where 152,050 square feet of space—including the Kashgar Grand Ballroom and six junior ballrooms—making it the ideal choice for everything from gala dinners and weddings to intimate meetings and international conferences. Further the dedicated events team is always on hand to design and execute the perfect, bespoke events package, complete with state-of-the-art audiovisual technology.

■ 客房及套房

酒店设有 3 896 间装潢雅致的客房及套房，均配备独有的喜来登甜梦之床（Sweet Sleeper™）。

酒店内全球规模最大的喜来登行政俱乐部，提供 570 间客房及套房，以及设有 274 个座位的喜来登行政酒廊。备有路氹景观客房、相连客房。房内配备 iPod 底座连收音机闹钟、高速上网、液晶电视机及有线频道、国际直拨及语音信箱功能电话、迷你酒吧、茶及咖啡冲调设备、独家提供喜来登炫逸水疗的沐浴用品、喜来登甜梦之床（Sweet Sleeper™）、摺叠床及婴儿床等可供选择。

■ Rooms and Suites

The hotel occupies 3,896 decorated guest rooms and suites with signature Sheraton Sweet Sleeper™ beds. The largest Sheraton Club Room around the world is set here, comprising of 570 guest rooms and suites, and Sheraton Club Lounge with 274 seats. The hotel also has Cotai View Rooms, Connected Rooms. The hotel offers in the rooms Ipod dock, clock radio, high-speed Internet access, LCD TV with cable channels, phones with functions of international direct dial and voicemails, mini bar, tea and coffee making facilities, Sheraton™ bath amenities, Sheraton Sweet Sleeper™ beds, rollaway beds and cribs.

OAKWORKS

shine
for Sheraton
喜来登水疗
spa

Le Méridien Coimbatore

哥印拜陀艾美酒店

Keywords 关键词
Decorative Design 装饰设计
Chakra Art 脉轮艺术
Excellent Service 卓越服务

酒店地址：印度泰米尔纳德邦，哥印拜陀 Avinashi 路 76 号，Neelambur 村
电　话：+ (91) (422) 2364343

Address: 762 Avinashi Road, Neelambur VillageCoimbatore, Tamil Nadu, India
Tel: + (91) (422) 2364343

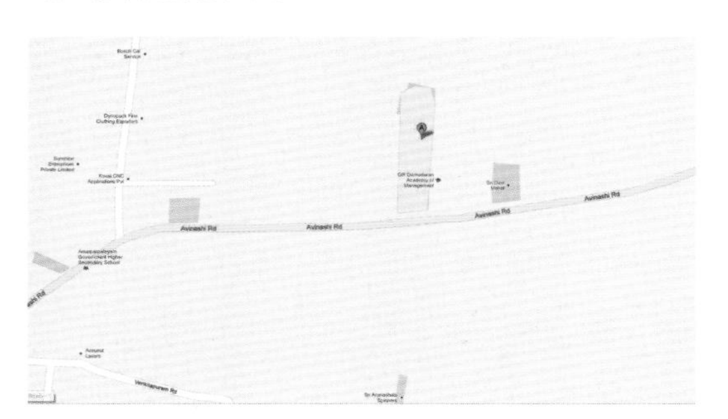

品牌链接

艾美酒店（Le Méridien），又被称为美丽殿酒店，是一家跨国的酒店品牌，原总部设于英国，属于喜达屋酒店及度假酒店国际集团。艾美酒店在全球 50 多个国家中有超过 120 个分店，主要设于欧洲、非洲、中亚、亚太地区和美国等地的热门旅游景点附近。

1972 年，法国航空建立了 Le Méridien 这个酒店品牌，以“提供客户宾至如归的感受”为目标。第一家艾美酒店（Le Méridien Etoile）于巴黎开幕，共有 1000 个房间，2 年内，共有 10 个分店纷纷于欧洲和非洲开幕。美丽殿酒店营运的前六年，开幕的酒店数目提高到 21 家，除了欧洲和非洲外，据点也拓展到中美洲、加拿大、南美洲、中亚等地。到了 1991 年中，艾美酒店的总数提升到 58 家。1994 年底，艾美酒店被英国的福特集团（Forte Group）买下，1996 年又被 Granada Group plc 收购，而福特集团的母公司格拉纳达集团（Granada Group plc）和康柏司集团（Compass Group plc）于 2000 年夏天合并，又在 2001 年 2 月分家，福特集团原有的三家酒店品牌，最后全归 Compass Group plc 所有。2001 年 5 月，野村财阀宣布以 19 亿英磅买下，并于 2001 年 2 月将艾美酒店与 Principal 酒店合并。2003 年 12 月，雷曼兄弟控股公司收购了艾美酒店。2005 年 11 月 24 日，艾美酒店和相关企业被喜达屋酒店及度假酒店国际集团并购。

About Le Méridien

Le Méridien Hotel is an international hotel brand and its original headquarters is in the UK, which belongs to the Starwood Hotels & Resorts Worldwide, Inc. Le Méridien Munich Hotel owns more than 120 branches in more than 50 countries around the world, mainly located near the popular tourist attractions in Europe, Africa, Central Asia, the Asia-Pacific region and the United States.

The Le Méridien brand was established in 1972 by Air France “to provide on a home away from home for its customers.” The first Le Méridien property was a 1,000-room hotel in Paris— Le Méridien Etoile. Within two years of operation, the group had 10 hotels in Europe and Africa. Within the first six years, the number of hotels had risen to 21 hotels in Europe, Africa, the French West Indies, Canada, South America, the Middle East and Mauritius. By 1991, the total number of Le Méridien properties had risen to 58. In late 1994, Le Méridien was acquired by UK hotel company Forte Group, which in turn was acquired by Granada plc in 1996. Through a merger in the summer of 2000 between Granada Group and global contract caterer Compass Group — and the subsequent de-merger of the two companies in February 2001— the ownership of the Forte Hotels division and its three brands (Le Méridien, Heritage Hotels and Posthouse Forte) passed solely to Compass. In May 2001, Nomura Group announced the acquisition of Le M é ridien Hotels & Resorts from Compass Group plc for £1.9 billion and Le Méridien was merged with Principal Hotels, which was acquired in February 2001. In December 2003, Lehman Brothers Holdings acquired the senior debt of Le Méridien. On November 24, 2005, the Le Méridien brand and management fee business was acquired by Starwood Hotels & Resorts. The leased and owned real estate assets were acquired in a separate deal by a joint venture formed by Lehman Brothers and Starwood Capital. In 2011, Le Méridien opened up its 100th hotel in Coimbatore, India. In mid 2012, during the search conducted by Maradu Municipality, the food safety officials seized damaged food from star hotels including Le Méridien Kochi.

MERIDIEN

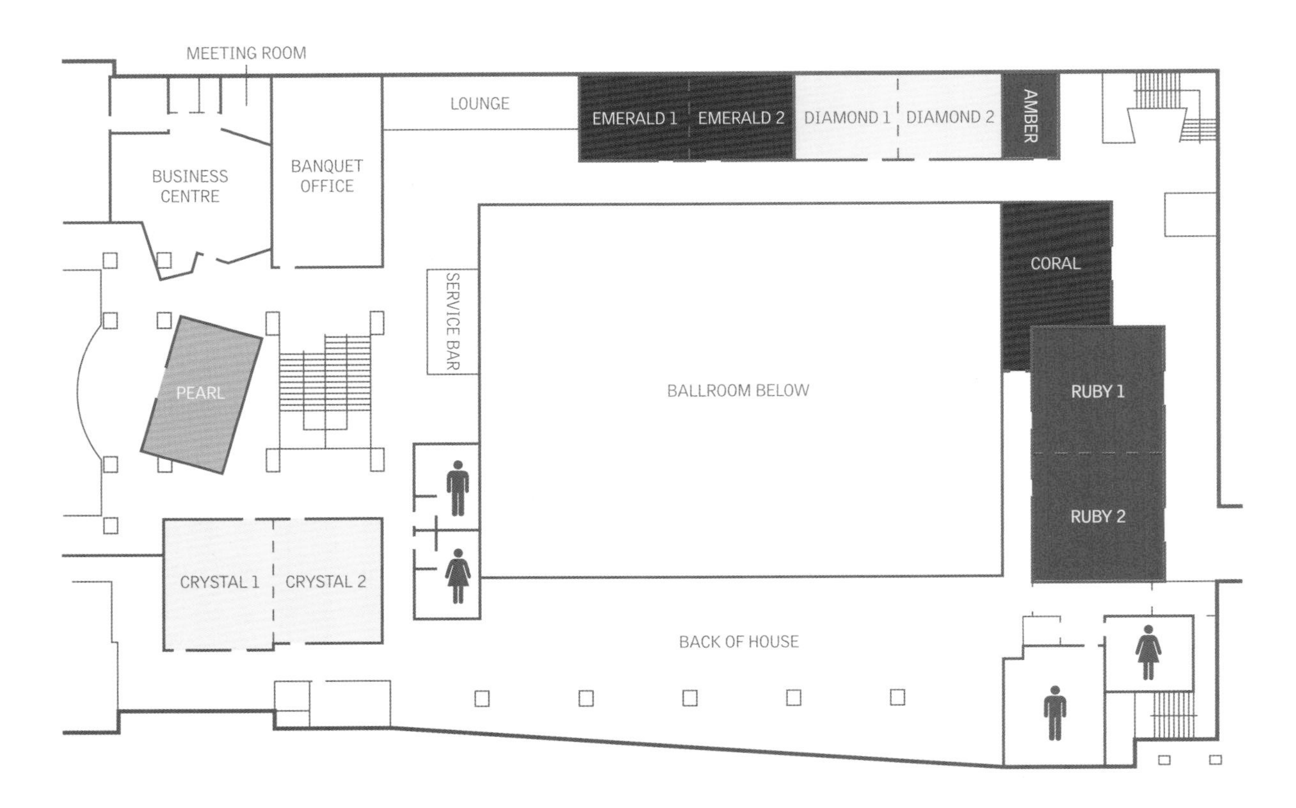
MEETING ROOM
LOUNGE
EMERALD 1
EMERALD 2
DIAMOND 1
DIAMOND 2
AMBER
BUSINESS CENTRE
BANQUET OFFICE
CORAL
SERVICE BAR
PEARL
BALLROOM BELOW
RUBY 1
RUBY 2
CRYSTAL 1
CRYSTAL 2
BACK OF HOUSE

酒店概况

哥印拜陀艾美酒店位于印度最具活力的城市之一，地理位置十分优越，紧邻哥印拜陀机场（CJB）、主要商务区和新建的资讯科技园。必将成为专业商旅人士的理想下榻之选。

Overview

Strategically positioned in one of India's most dynamic cities, Le Méridien Coimbatore is close to Coimbatore Airport (CJB), major businesses, and a new IT park. It's an ideal space for the sophisticated business professional.

酒店特色

艾美酒店因其热情周到的宾客服务和对细节的关注而享誉全球。在体验艾美国际化特色与活动的同时，顾客还会全方位领略到其无可挑剔的卓越服务。

Feature

Le Méridien is worldly renowned for its attention to detail and dedication to exemplary customer service. In addition to all of the global features and activities people would expect at Le Méridien Hotels and Resorts, impeccable service is promised at every turn.

酒店室内

酒店的 254 间客房与套房环境舒适、技术先进，处处彰显出鲜明的个性与创造力。每间客房均拥有迷人的装饰设计和灵活的工作空间。温馨的泥土色调与豪华的起居区及办公区相得益彰，共同打造令人陶醉的舒适氛围。

Interior

With an emphasis on comfort and technology, 254 guest rooms and suites evoke individuality and creativity. Featuring engaging décor and flexible work space, the rooms offer an urban escape for the well-travelled.

酒店配套

■ 餐饮

致力于提供美味佳肴和香醇佳酿是艾美酒店始终如一的服务宗旨及传统。拥有各类大小餐厅，可尽情品尝可口美食。

Services and Amenities

■ Dining

Dedication to offering delicious cuisine and fine wine is an integral part of the heritage and tradition of Le Méridien. Savour delicious gastronomy elegantly presented at each of the dining venues.

■ 庆典及会议

哥印拜陀艾美酒店拥有多种会议室以及面积近 4 300 m^2 的现代化活动空间，其中包括一间 1 194 m^2 的宴会厅。先进一流的视听技术、专业齐全的商务中心和创意十足的工作人员必将确保顾客在这里举办的每次会议与活动都能激发灵感、成功难忘。

■ 脉轮艺术

绚丽多彩的药泥为生活带来了七种颜色的脉轮，从而将身体打造成为了一个活生生的艺术品。客人可以通过温暖床单裹敷尽享舒适惬意，而排毒药泥则可以舒缓各种疼痛和软化肌肤。使用热毛巾将药泥取出，为身体保湿按摩做准备。

■ Celebrations and Other Events

Le Méridien Coimbatore offers a variety of meeting rooms and almost 4,300 square metres of contemporary function space, including a 1,194-square-metre ballroom. State-of-the-art audiovisual technology, a professional business centre, and creative staff members guarantee that every meeting hosted here will be an inspiring and successful event.

■ Chakra Art

Colour-infused therapeutic muds bring the seven colours of the chakras to life making the body into a living work of art. Guests are wrapped in warm linen sheets while the detoxifying mud soothes aches and pains and softens the skin. The mud is removed with hot towels, preparing the body for a moisturizing massage.

இருந்தோம்பி இல்வாழ்வ தெல்லாம் விருந்தோம்பி
வேளாண்மை செய்தற் பொருட்டு.

இருந்தோம்பி இல்வாழ்வ

图书在版编目（CIP）数据

酒店+2 奢华风 : 汉英对照 / 佳图文化 主编. -- 北京 : 中国林业出版社, 2013.7

ISBN 978-7-5038-7074-3

Ⅰ. ①酒… Ⅱ. ①佳… Ⅲ. ①建筑设计—图集 ②景观设计—图集 Ⅳ. ① TU206 ② TU986.2-64

中国版本图书馆 CIP 数据核字 (2013) 第 299823 号

策　　划：王　志
主　　编：佳图文化

中国林业出版社·建筑与家居图书出版中心
责任编辑：李　顺　　唐　杨
出版咨询：（010）83223051

出 版：中国林业出版社（100009　北京西城区德内大街刘海胡同 7 号）
网 站：http://lycb.forestry.gov.cn/
印 刷：利丰雅高印刷（深圳）有限公司
发 行：中国林业出版社
电 话：（010）83224477
版 次：2013 年 7 月第 1 版
印 次：2013 年 7 月第 1 次
开 本：889mm×1194mm　1/16
印 张：18
字 数：200 千字
定 价：328.00 元（USD 60.00）